International Review of Science

Organic Chemistry
Series Two

Consultant Editor
D. H. Hey, F.R.S.

Publisher's Note

The International Review of Science is an important venture in scientific publishing presented by Butterworths. The basic concept of the Review is to provide regular authoritative reviews of entire disciplines. Chemistry was taken first as the problems of literature survey are probably more acute in this subject than in any other. Biochemistry and Physiology followed naturally. As a matter of policy, the authorship of the Review of Science is international and distinguished, the subject coverage is extensive, systematic and critical.

The Review has been conceived within a carefully organised editorial framework. The overall plan was drawn up and the volume editors appointed by seven consultant editors. In turn, each volume editor planned the coverage of his field and appointed authors to write on subjects which were within the area of their own research experience. No geographical restriction was imposed. Hence the 500 or so contributions to the Review of Science come from many countries of the world and provide an authoritative account of progress.

The publication of Organic Chemistry Series One was completed in 1973 with ten text volumes and one index volume; in accordance with the stated policy of issuing regular reviews to keep the series up to date, volumes of Series Two will be published between the middle of 1975 and early 1976; Series Two of Physical Chemistry will be published at the same time, while Inorganic Chemistry Series Two was published during the first half of 1975. Volume titles are the same as in Series One but the articles themselves either cover recent advances in the same subject or deal with a different aspect of the main theme of the volume. In Series Two an index is incorporated in each volume and there is no separate index volume.

Butterworth & Co. (Publishers) Ltd.

ORGANIC CHEMISTRY SERIES TWO

Consultant Editor
D. H. Hey, F.R.S., *formerly of the Department of Chemistry, King's College, University of London*

Volume titles and Editors

1 STRUCTURE DETERMINATION IN ORGANIC CHEMISTRY
Professor L. M. Jackman, *Pennsylvania State University*

2 ALIPHATIC COMPOUNDS
Professor N. B. Chapman, *University of Hull*

3 AROMATIC COMPOUNDS
Professor H. Zollinger, *Eidgenössische Technische Hochschule, Zürich*

4 HETEROCYCLIC COMPOUNDS
Professor K. Schofield, *University of Exeter*

5 ALICYCLIC COMPOUNDS
Professor D. Ginsburg, *Technion-Israel Institute of Technology, Haifa*

6 AMINO ACIDS, PEPTIDES AND RELATED COMPOUNDS
Professor H. N. Rydon, *University of Exeter*

7 CARBOHYDRATES
Professor G. O. Aspinall, *York University, Ontario*

8 STEROIDS
Dr. W. F. Johns, *G. D. Searle & Co., Chicago*

9 ALKALOIDS
Professor K. Wiesner, F.R.S., *University of New Brunswick*

10 FREE RADICAL REACTIONS
Professor W. A. Waters, F.R.S., *University of Oxford*

INORGANIC CHEMISTRY SERIES TWO

Consultant Editor
H. J. Emeléus, C.B.E., F.R.S. *Department of Chemistry University of Cambridge*

Volume titles and Editors

1 MAIN GROUP ELEMENTS—HYDROGEN AND GROUPS I–III
Professor M. F. Lappert, *University of Sussex*

2 MAIN GROUP ELEMENTS—GROUPS IV AND V
Dr. D. B. Sowerby, *University of Nottingham*

3 MAIN GROUP ELEMENTS—GROUPS VI AND VII
Professor V. Gutmann, *Technical University of Vienna*

4 ORGANOMETALLIC DERIVATIVES OF THE MAIN GROUP ELEMENTS
Professor B. J. Aylett, *Westfield College, University of London*

5 TRANSITION METALS— PART 1
Professor D. W. A. Sharp, *University of Glasgow*

6 TRANSITION METALS— PART 2
Dr. M. J. Mays, *University of Cambridge*

7 LANTHANIDES AND ACTINIDES
Professor K. W. Bagnall, *University of Manchester*

8 RADIOCHEMISTRY
Dr. A. G. Maddock, *University of Cambridge*

9 REACTION MECHANISMS IN INORGANIC CHEMISTRY
Professor M. L. Tobe, *University College, University of London*

10 SOLID STATE CHEMISTRY
Dr. L. E. J. Roberts, *Atomic Energy Research Establishment. Harwell*

PHYSICAL CHEMISTRY SERIES TWO

Consultant Editor
A. D. Buckingham, F.R.S., *Department of Chemistry University of Cambridge*

Volume titles and Editors

1 THEORETICAL CHEMISTRY
Professor A. D. Buckingham, F.R.S., *University of Cambridge* and Professor C. A. Coulson, F.R.S., *University of Oxford*

2 MOLECULAR STRUCTURE AND PROPERTIES
Professor A. D. Buckingham, F.R.S., *University of Cambridge*

3 SPECTROSCOPY
Dr. D. A. Ramsay, F.R.S.C., *National Research Council of Canada*

4 MAGNETIC RESONANCE
Professor C. A. McDowell, F.R.S.C., *University of British Columbia*

5 MASS SPECTROMETRY
Professor A. Maccoll, *University College, University of London*

6 ELECTROCHEMISTRY
Professor J. O'M. Bockris, *The Flinders University of S. Australia*

7 SURFACE CHEMISTRY AND COLLOIDS
Professor M. Kerker, *Clarkson College of Technology, New York*

8 MACROMOLECULAR SCIENCE
Professor C. E. H. Bawn, C.B.E., F.R.S., *formerly of the University of Liverpool*

9 CHEMICAL KINETICS
Professor D. R. Herschbach *Harvard University*

10 THERMOCHEMISTRY AND THERMO-DYNAMICS
Dr. H. A. Skinner, *University of Manchester*

11 CHEMICAL CRYSTALLOGRAPHY
Professor J. M. Robertson, C.B.E., F.R.S., *formerly of the University of Glasgow*

12 ANALYTICAL CHEMISTRY —PART 1
Professor T. S. West, *Macaulay Institute for Soil Research, Aberdeen*

13 ANALYTICAL CHEMISTRY —PART 2
Professor T. S. West, *Macaulay Institute for Soil Research, Aberdeen*

BIOCHEMISTRY
SERIES ONE

Consultant Editors
H. L. Kornberg, F.R.S.
Department of Biochemistry
University of Leicester and
D. C. Phillips, F.R.S., *Department of*
Zoology, University of Oxford

Volume titles and Editors

1 CHEMISTRY OF MACRO-
MOLECULES
Professor H. Gutfreund, *University of*
Bristol

2 BIOCHEMISTRY OF CELL WALLS
AND MEMBRANES
Dr. C. F. Fox, *University of California,*
Los Angeles

3 ENERGY TRANSDUCING
MECHANISMS
Professor E. Racker, *Cornell University*
New York

4 BIOCHEMISTRY OF LIPIDS
Professor T. W. Goodwin, F.R.S.,
University of Liverpool

5 BIOCHEMISTRY OF CARBO-
HYDRATES
Professor W. J. Whelan, *University*
of Miami

6 BIOCHEMISTRY OF NUCLEIC
ACIDS
Professor K. Burton, F.R.S., *University of*
Newcastle upon Tyne

7 SYNTHESIS OF AMINO ACIDS
AND PROTEINS
Professor H. R. V. Arnstein, *King's*
College, University of London

8 BIOCHEMISTRY OF HORMONES
Professor H. V. Rickenberg, *National*
Jewish Hospital & Research Center,
Colorado

9 BIOCHEMISTRY OF CELL DIFFER-
ENTIATION
Professor J. Paul, *The Beatson Institute*
for Cancer Research, Glasgow

10 DEFENCE AND RECOGNITION
Professor R. R. Porter, F.R.S., *University*
of Oxford

11 PLANT BIOCHEMISTRY
Professor D. H. Northcote, F.R.S.
University of Cambridge

12 PHYSIOLOGICAL AND PHARMACO-
LOGICAL BIOCHEMISTRY
Dr. H. K. F. Blaschko, F.R.S., *University*
of Oxford

PHYSIOLOGY
SERIES ONE

Consultant Editors
A. C. Guyton,
Department of Physiology and
Biophysics, University of Mississippi
Medical Center and
D. F. Horrobin,
Department of Physiology, University
of Newcastle upon Tyne

Volume titles and Editors

1 CARDIOVASCULAR PHYSIOLOGY
Professor A. C. Guyton and Dr. C. E. Jones,
University of Mississippi Medical Center

2 RESPIRATORY PHYSIOLOGY
Professor J. G. Widdicombe, *St. George's*
Hospital, London

3 NEUROPHYSIOLOGY
Professor C. C. Hunt, *Washington*
University School of Medicine, St. Louis

4 GASTROINTESTINAL PHYSIOLOGY
Professor E. D. Jacobson and Dr. L. L.
Shanbour, *University of Texas Medical*
School

5 ENDOCRINE PHYSIOLOGY
Professor S. M. McCann, *University of*
Texas

6 KIDNEY AND URINARY TRACT
PHYSIOLOGY
Professor K. Thurau, *University of Munich*

7 ENVIRONMENTAL PHYSIOLOGY
Professor D. Robertshaw, *University*
of Nairobi

8 REPRODUCTIVE PHYSIOLOGY
Professor R. O. Greep, *Harvard Medical*
School

International Review of Science

Organic Chemistry
Series Two

Volume 1
Structure Determination in Organic Chemistry

Edited by **L. M. Jackman,**
Pennsylvania State University

BUTTERWORTHS
LONDON - BOSTON
Sydney - Wellington - Durban - Toronto

THE BUTTERWORTH GROUP

ENGLAND
Butterworth & Co (Publishers) Ltd
London: 88 Kingsway, WC2B 6AB

AUSTRALIA
Butterworths Pty Ltd
Sydney: 586 Pacific Highway, NSW 2067
Also at Melbourne, Brisbane, Adelaide and Perth

CANADA
Butterworth & Co (Canada) Ltd
Toronto: 2265 Midland Avenue, Scarborough, Ontario M1P 4S1

NEW ZEALAND
Butterworths of New Zealand Ltd
Wellington: 26–28 Waring Taylor Street, 1

SOUTH AFRICA
Butterworth & Co (South Africa) (Pty) Ltd
Durban: 152–154 Gale Street

USA
Butterworths (Publishers) Inc
Boston: 19 Cummings Park, Woburn, Mass., 01801

Library of Congress Cataloging in Publication Data

Main entry under title:
Structure determination in organic chemistry.

(Organic chemistry, series two; v. 1) (International
review of science)
   Includes index.
   1. Chemistry, Organic.   2. Chemistry, Analytic.
I. Jackman, Lloyd Miles.   II. Series.   III. Series:
International review of science.
QD245.073   vol. 1 [QD271]   547'.008s   [547'.1'22]
ISBN 0 408 70613 9   75-19449

First Published 1976 and © 1976
BUTTERWORTH & CO (PUBLISHERS) LTD

Typeset, printed and bound in Great Britain by
REDWOOD BURN LIMITED
Trowbridge & Esher

Consultant Editor's Note

The ten volumes in Organic Chemistry in the Second Series of the biennial reviews in the International Review of Science follow logically from those of the First Series. No major omissions have come to light in the overall coverage of the First Series. The titles of the ten volumes therefore remain unchanged but there are three new Volume Editors. The volume on Structure Determination in Organic Chemistry has been taken over by Professor Lloyd M. Jackman of Pennsylvania State University, that on Alicyclic Compounds by Professor D. Ginsburg of Technion-Israel Institute of Technology, and that on Amino Acids, Peptides and Related Compounds by Professor H. N. Rydon of the University of Exeter. The international character of the Series is thus strengthened with four Volume Editors from the United Kingdom, two each from Canada and the U.S.A., and one each from Israel and Switzerland. An even wider pattern is shown for the authors, who now come from some sixteen countries. The reviews in the Second Series are mainly intended to cover work published in the years 1972 and 1973, although relevant results published in 1974 and 1975 are included in some cases, and earlier work is also covered where applicable.

It is my pleasure once again to thank all the Volume Editors for their helpful cooperation in this venture.

London   D. H. Hey

Preface

The determination of structure and stereochemistry is fundamental to all areas of organic chemistry and often assumes a crucial role in biochemistry. The synthetic organic chemist, for instance, is usually confronted with a structural problem at each step in a synthesis. Similarly, much of modern physical organic chemistry involves the prediction and subsequent verification of structures of products formed under various conditions. The process of structure determination may vary from a casual examination of, say, an n.m.r. spectrum to a highly refined X-ray crystallographic analysis. One point is clear, however, and that is that determination of structure is involving a decreasing dependence on chemical degradative methods which were the corner stones of structure determinations up to the 1950s. Today, we find that problems are usually solved rapidly by the applications of spectroscopic methods and a minimal number of chemical transformations or, for reasons of complexity or subtlety, merit the use of X-ray crystallography.

Spectroscopic methods, particularly those reviewed in Chapters 1, 2 and 3, are rapid and relatively inexpensive. They are generally most effective if partial structural information is already available, as is often the case with those problems which arise in the course of a multistep synthesis, or in the investigation of a biogenetically related group of natural products. In addition, these methods are applicable to all phases of matter and may therefore be used for studying ephemeral species such as ion pairs, weak complexes, etc. which exist in solution, but not necessarily in the crystalline state. Furthermore, spectroscopic methods frequently provide information concerning inter- and intra-molecular dynamics.

X-ray crystallography (Chapter 4) has just entered a new phase in its development. It is now a commercially available service with charges for the work quotable in advance. The total cost is still high, but in many instances, where crystalline compounds are under consideration, cost effectiveness is favourable. Furthermore, the method can rather readily be refined to afford absolute configurations of enantiomers as well as their structures and relative stereochemistries.

Structural organic chemistry in its own right has usually been synonomous with natural product chemistry. Activity in this latter area has been steadily increasing and is an appropriate subject for review (Chapter 5) in this volume, together with recent studies of biosyntheses, since the natural product chemist makes considerable use of biosynthetic arguments in postulating plausible structures which can be subjected to experimental verification.

Pennsylvania   L. M. Jackman

Contents

Mass spectrometry   1
J. M. Wilson, *University of Manchester*

Ultraviolet and visible spectroscopy   35
C. J. Timmons, *University of Nottingham*

Nuclear magnetic resonance spectroscopy   55
I. O. Sutherland, *University of Sheffield*

X-ray crystallography   99
A. F. Cameron, *University of Glasgow*

Natural products — structure determination   125
E. Haslam, *University of Sheffield*

Index   165

1
Mass Spectrometry

J. M. WILSON
University of Manchester

1.1   INTRODUCTION   1

1.2   INSTRUMENTAL METHODS   2
      1.2.1   *Chemical ionisation*   2
      1.2.2   *Field ionisation*   4
      1.2.3   *Photoionisation*   6
      1.2.4   *Laser ionisation*   7
      1.2.5   *Negative ion production*   7
      1.2.6   *Data acquisition and handling*   8
      1.2.7   *Techniques for investigating ion structure*   10

1.3   INVESTIGATION OF ION STRUCTURES   12
      1.3.1   *Aromatic species*   12
      1.3.2   *Aliphatic species*   15
      1.3.3   *Doubly charged ions*   16

1.4   ION–MOLECULE REACTIONS   18
      1.4.1   *Gas-phase acid–base relationships*   18
      1.4.2   *Gas-phase solvation of ions*   21
      1.4.3   *Other ion–molecule reactions*   21

1.5   DETERMINATION OF MOLECULAR STRUCTURE   22
      1.5.1   *Amino acids and peptides*   22
      1.5.2   *Other natural products*   26
      1.5.3   *Stereochemical problems*   27

1.1   INTRODUCTION

Since the last article in this series was written, a number of the instrumental developments described have passed into more common use. GC–MS is now used routinely in many laboratories, and data systems, field ionisation (FI) and chemical ionisation (CI) sources are now offered together by instrument manufacturers as a standard package (though not a cheap one). In the section on instrumental methods, new developments are discussed, but general applications of the new ionisation methods will be found in the sections on applications.

Ion cyclotron resonance (ICR) remains an important research technique and has been used to obtain many of the results on ion–molecule reactions. The complementary results which are being obtained, on the one hand, from acid and base strengths in the gas phase and, on the other, from the behaviour of isolated solvated ions in the gas phase should lead to a better understanding of the properties of ions in solution.

There is not enough space for this review article to cover comprehensively the detailed behaviour of various types of organic compound in the mass spectrometer, even if the complete chapter were devoted to this aspect. The emphasis has been placed on applications which stretch the method to its current limits or show where these might be extended.

1.2 INSTRUMENTAL METHODS

1.2.1 Chemical ionisation

The situation required for chemical ionisation spectra, i.e. a sample at a very low partial pressure in the presence of a much greater pressure of reactant gas, lends itself to combination with gas chromatography[1]. Attempts have more recently been made to extend this to combination with liquid chromatography. The method described[2] does not involve a direct link, but is a direct analysis of batches of dilute solutions. A liquid flow through a glass capillary of 0.01 ml min^{-1} is sufficient to maintain CI conditions in the ion source, using the evaporated solvent as reactant gas. Ammonia, water, methanol and hexane have all been used successfully. They are not all ideal reactant gases in that methanol forms protonated oligomers up to $(MeOH)_5H^+$ which may interfere with sample ions. Hexane does not present this problem, having no ions heavier than $C_6H_{13}^+$, but it will not ionise saturated hydrocarbons. This application is not yet fully realised.

'Direct chemical ionisation' is the name applied to the technique in which the solid sample is directly exposed to the ion plasma in the ion source[3]. This allows analysis of samples at much lower temperatures than have previously been possible, and of much less volatile samples. Using this method, free tripeptides produce CI spectra essentially free from decomposition and at its limits a skilled operator can obtain spectra from peptides such as $(Ala)_6$ and H_2N-Pro-Phe-His-Leu-Leu-OH.

Good yields of negative ions can be obtained from chloro compounds using methane CI[4]. In such cases the reactant ions are Cl^-, H_2OCl^- and HCl_2^- from the compound itself and residual water in the instrument rather than from the reactant gas. The ion yield was less using CH_2Cl_2 as reactant gas.

An atmospheric pressure ion source has been described in which very high sensitivities have been achieved for both positive and negative ions[5]. Initial ionisation is by radiation from a ^{63}Ni foil. When nitrogen containing small amounts of benzene is introduced it reacts by charge transfer to most molecules:

$$C_6H_6^+\cdot + M \rightarrow M^{+\cdot} + C_6H_6$$

but proton transfer occurs with more basic molecules:

$$C_6H_6^+\cdot + M \rightarrow MH^+ + C_6H_5\cdot$$

In the presence of chloroform, barbiturate molecules BH form B^- ions, probably by reaction with Cl^-. The limit of detection for positive ion production in this source is in the 5–10 pg region for single ion monitoring and in the 100 pg region for 0–750 amu scanning.

In chemical ionisation sources cooled with liquid nitrogen it is possible to observe condensation ions which are held together by ion-induced dipole forces[6]. In H_2 the following equilibria are observed:

$$H_3^+ + H_2 \rightleftharpoons H_5^+$$
$$H_5^+ + H_2 \rightleftharpoons H_7^+$$

The enthalpy change for the H_5/H_7 equilibrium is only -1.8 kcal mol^{-1}.

The reactions of methane-derived ions under CI conditions can also be observed at lower pressures using ion cyclotron resonance. Ejection experiments show that CH_5^+ and $C_2H_5^+$ contribute to all the ions in the hexane CI spectrum, with the exception of $C_5H_{11}^+$ [7]. This would suggest that only CH_5^+ is a sufficiently strong acid to protonate a methyl group. Chemical ionisation has also been observed at lower pressures using a three-dimensional quadrupole ion storage trap (Quistor)[8]. The production of fragment ions from saturated hydrocarbons by methane CI is more complicated than was originally assumed[9]. Scrambling of deuterium and hydrogen atoms in lower mass $C_nH_{2n+1}^+$ ions in the spectra of specifically deuterated n-decane is interpreted to mean that the principal route to these ions is

$$C_{10}H_{22} \xrightarrow[C_2H_5^+]{CH_5^+} C_{10}H_{21}^+ \rightarrow C_nH_{2n+1}^+ + C_{10-n}H_{20-2n}$$

the randomisation taking place during or prior to unimolecular decomposition of $C_{10}H_{21}^+$. This two-step process may be accompanied by a single-step process involving electrophilic attack. The latter is the only process which can form $C_9H_{19}^+$. There is further evidence for unimolecular decompositions in methane CI presented in a study of model compounds related to glycosides[10]. A prominent ion in the spectrum of compound (1) is a fragment which is assigned the structure (2). The H:D ratio in (2) suggests that it must be formed

(1)   (2)

by a simple rearrangement process from the protonated molecular ion. In the methane CI spectra of alkyl phenyl ketones[11] the variation in the $[M + H]^+$: $[M - H]^+$ ratio with chain length is consistent with the suggestion that $[M - H]^+$ formation is a bimolecular hydride abstraction process.

Much useful effort has gone into the search for alternative reactant gases in CI. Mixed gases can be extremely useful for GC–CI–MS work, if the major component is an inert gas also used as carrier gas and the second component can be mixed with the column effluent. Standard conditions can therefore be maintained in the solumn. An He/H_2O mixture[12] gives a spectrum in which there are abundant fragment ions, produced by dissociative charge exchange from He^+, and usually an abundant $[M + H]^+$ ion produced by protonation from H_3O^+. Ammonia may be a useful reactant gas. The NH_4^+ ion is a much weaker acid than the carbonium and carbenium ions and it protonates selectively. α,β-Unsaturated ketones are protonated but not saturated ketones, although most oxygen compounds do form condensation ions MNH_4^+ [13].

Nitric oxide is also selective in the ionisation processes observed. Ketones, esters and carboxylic acids all produce $[M + NO]^+$ ions whereas aldehydes and ethers produce principally $[M - H]^+$ ions[14]. It also ionises by charge exchange and, in a study of a selection of gases, NO and CO were found to give the best charge-exchange spectra for organic purposes[15]. A study of the charge-exchange spectra of substituted benzophenones showed in all cases reasonably linear Hammett plots for the $[RC_6H_4CO^+]:[C_6H_5CO^+]$ ratio. The smaller the recombination energy of the charge exchange gas, the greater is the substituent effect[16].

1.2.2 Field ionisation and field desorption

The developments made in this type of source have been aimed at improving performance and extending the range of compound types which can be used. A considerable amount of work has gone into the determination of optimum parameters for field ion emitter wires[17]. The principle parameters are emitter temperature, activation time, position of emitter and emitter potential. Activation conditions are extremely important since this determines the nature of the microneedles on the surface of the wire. Emitters activated at high temperatures have microneedles which are stable to chemical attack by dilute mineral acid, alkali and gaseous fluorocarbons[18]. Purine bases deposited on such emitters in HCl solution gave much more intense spectra than those deposited from aqueous solution. Guanine deposited from dilute NaOH solution gave a spectrum in which the principal ions were M^+, MH^+ and MNa^+. The principal ion in the field desorption (FD) mass spectrum of sodium acetate is Na_2OAc^+. The usual procedure in running FD spectra of involatile materials is to raise the emitter temperature by passing a current through it. This is thought to produce a non-uniform temperature distribution, particularly in the microneedles. The alternative method of indirectly heating the emitter with near-infrared radiation leads to spectra with more abundant molecular ions and less fragmentation[19]. The method of deposition of samples from solution makes the method attractive in combination with liquid chromatography[20].

In the field ionisation spectra of a series of pentaerythritol derivatives (3; X = OH or Br) the $[M + H^+]:[M^+]$ ratio increases with increasing number

$$XCH_2 \overset{\overset{\displaystyle CH_2X}{|}}{\underset{\underset{\displaystyle CH_2X}{|}}{C}} CH_2OH$$

(3)

of hydroxyl groups[21]. The explanation given is that the density of adsorbed molecules at the emitter surface will be greater for more polar compounds. In mixtures of polar and non-polar compounds, preferential adsorption of polar compounds partially suppresses the ionisation of the less polar compounds and results in a non-linear sensitivity response.

In the field ionisation of substituted acetophenones the process (4) → (5) can be observed:

$$RC_6H_4COCH_3^{+\cdot} \rightarrow RC_6H_4CO^+ + CH_3\cdot$$
$$\quad\text{(4)} \qquad\qquad\qquad \text{(5)}$$

The (4):(5) ratio does not give a linear correlation with Hammett constants for 'normal' fragments formed at the emitter surface, but does so for meta-stable ions[22]. The latter follow the same kinetic equations as ions formed by electron impact, but field-induced dissociation at the surface will be very much dependent upon the orientation of the adsorbed molecules towards the surface and this is a property which cannot be easily predicted for different substituents.

The kinetics of unimolecular decomposition in FI can be studied for very fast rates (in the 10^{10} s^{-1} region) either by analysis of the 'fast' metastable ions on the low mass tail of normal fragments in a single-focusing mass spectrometer, or by changing the emitter potential in a double-focusing instrument. The first method was used to show that, in the spectrum of menthone (6),

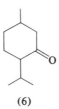

(6)

decomposition rates increase with increasing thermal energy. $[M - C_3H_6]^{+\cdot}$, a rearrangement ion, has a lower activation energy than $[M - CH_3]^+$ which is probably formed by a simple fission process, but the curves cross at higher thermal energies[23]. In general, skeletal rearrangements are slow and hydrogen rearrangements fast on the FI timescale, but some direct bond cleavages are also slow, so a definite distinction cannot always be drawn[24].

A number of interesting kinetic analyses have been made, by FI in a double-focusing mass spectrometer, of processes which have been much studied under electron impact. The loss of C_2H_4 from the molecular ion of cyclohexene has been shown to be a retro-Diels–Alder reaction[25]. At 10^{-10} s the predominant species in the FI spectrum of cyclo[3,3,6,6-$^2H_4$]hexene is $C_4H_2D_4^+\cdot$, but H/D randomisation is complete by 10^{-9} s. The two possible products, (8) and (9), of McLafferty rearrangement of [4,4-$^2H_2$]hexanal (7) have different rate profiles, (8) predominating at short times and (9) predominating after longer

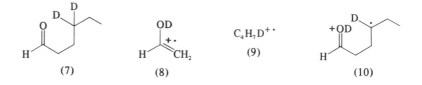

(7)   (8) $C_4H_7D^+\cdot$ (10)
                                        (9)

times[26]. The explanation given is that the slower process to form (9) involves a further H rearrangement in the intermediate (10) and that the structure of (9) is but-2-ene rather than but-1-ene. This is not inconsistent with the results of studies on labelled heptanal, where $C_2H_4O^+\cdot$ is formed without randomisation on any timescale but the formation of $C_5H_{10}^+\cdot$ is accompanied by considerable randomisation of hydrogen atoms under EI (electron ionisation) and to a reduced extent under FI[27]. The FI mass spectrum of $CD_3COCD_2C_5H_{11}$ shows no randomisation at times less than 7×10^{-10} s both for $CD_3C(OH)$-$CD_2^+\cdot$ and $[M - CD_3]^+$ formation[28]. In a similar study of the elimination of water from n-hexanol molecular ions, it was found that 1,4-elimination predominates in low-energy (slowly decomposing) ions whereas 1,3- and 1,4-processes compete effectively in the faster decomposing ions[29]. The 70 eV EI results (predominantly 1,4) are a result of the fact that the low-energy ions survive as $C_6H_{12}^+\cdot$ whereas the higher-energy ions decompose further to C_4 and C_5 ions.

1.2.3 Photoionisation

The use of Ar (11.6 eV), Kr (10.1 eV) and Xe (8.5 eV) discharge lamps as photon sources has been applied to the study of reaction intermediates, particularly radicals. This method has the advantage that at such low energies there should be little interference from fragment ions at the same mass as radicals and there are no problems of decomposition at the filament to produce a spurious source of radicals or of reaction of corrosive gases with the filament. Such a system using interchangeable lamps has been used to identify $C_4H_8O_3$ and several radical species as intermediates in the but-2-ene–ozone reaction[30]. In the reaction of oxygen atoms with acetylene, HCCO· and CH_2 have been detected. Steady-state concentrations of ca. 10^{-10} mol l^{-1} are detectable[31] in systems with a reaction pressure of 1 Torr.

The process $C_7H_8^+\cdot \rightarrow C_7H_7^+ + H\cdot$ in toluene is much faster under photon impact than when induced by electrons of the same energy[32]. Similar comparisons for dec-2-ene show that the molecular ion is less abundant in the PI spectrum than in the EI spectrum[33]. This is to some extent a result of the difference in energy deposition functions for the processes

$$M + h\nu \rightarrow M^{+}\cdot + e^{-}$$

and

$$M + e^{-} \rightarrow M^{+}\cdot + 2e^{-}$$

and the authors suggest that photoionisation spectra are not really 'simpler' than low-energy electron impact mass spectra. It does not necessarily follow that the same will be true of more polar compounds where the absence of a filament in the source region may improve spectra. Certainly advantages are claimed for photoionisation in work with alcohols[34] and peptide derivatives[35].

The study of single events in photoionisation should lead to a better understanding of ion chemistry. It is possible, for such an event, to measure simultaneously the kinetic energy of the ejected electron and the mass of the ion produced. The basic principles of PE–PI theory have been worked out[36] and a successful instrument produced[37]. Results with CH_2Cl_2 show that $CH_2Cl^{+}\cdot$ is produced from the ground state of the ion and CH_2Cl^{+} from the first excited state. An interesting application of the coincidence method is to the photoionisation spectrum of C_2F_6, in which are two abundant ions, CF_3^{+} with an onset of *ca.* 13.7 eV and $C_2H_5^{+}$ with an onset *ca.* 15.3 eV. Ions in the first excited state, at 16 eV, decompose mostly to $C_2F_5^{+}$ rather than by the lower energy process to CF_3^{+} [38]. This suggests that there is a direct decomposition of the first excited state of the molecular ion, which is not in quasi-equilibrium with the ground state. This is a direct contradiction of the quasi-equilibrium theory of mass spectra.

1.2.4 Laser ionisation

There are as yet very few examples of this type of ion source, using a laser for both volatilisation and ionisation of solid samples[39]. The nature of the ionisation process is not well understood. Spectra obtained from sodium alkylsulphonates using a 0.1 J, 6943 Å ruby laser show[40], as the predominant ion, $[M + Na]^{+}$. Sodium alkylthiosulphates behave similarly and also give inorganic fragments such as $Na_3SO_3^{+}$. The utility of the method has still to be explored and compared with field desorption, which can also yield spectra from salts[18].

1.2.5 Negative ion production

The production of negative ions can occur by three processes, electron capture

$$M + e^{-} \rightarrow M^{-}\cdot$$

dissociative electron capture

$$X—Y + e^{-} \rightarrow X^{-} + Y\cdot$$

and ion pair formation

$$X—Y + e^{-} \rightarrow X^{-} + Y^{+} + e^{-}$$

Most electron capture processes require electron energies in the region of 0–3 eV and ion pair formation occurs in the region 8–12 eV. The production of electron beams of less than 5 eV presents instrumental problems where there are high accelerating fields in the source region. This is not, however, a problem in ion cyclotron resonance spectrometers and a considerable amount of work on negative-ion–molecule reactions has been done and will be discussed in Section 1.3. In conventional mass spectrometers negative ion spectra have been obtained at 70 eV and these are almost certainly produced by capture of slow secondary electrons produced by positive ion formation. The ionisation efficiency curve of $C_6H_5NO_2^-\cdot$ has a similar shape[41] to that of $C_6H_5NO_2^+\cdot$ above 10 eV, and in the region 3–20 eV for the same compound the $M^-\cdot$ abundance increases while that of negative fragment ions decreases[42]. Negative ion mass spectra of nitroarenes contain fewer ionic species than do the positive ion spectra. In the nitroacenaphthenes, loss of NO predominates rather than loss of NO_2 as in the positive ion spectra. Requirements for loss of ·OH from the molecular negative ion are the same as in positive ions[43]. The fragmentation of negative ions of phosphoranes can be rationalised by the formation of trivalent phosphorus species, as in the fragmentation scheme shown for (11)[44].

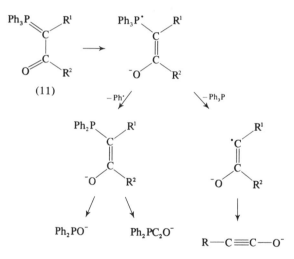

Sulphur compounds also give abundant $M^-\cdot$ ions[45]. Dicarboxylic acids generally exhibit $[M - H]^-$ ions, more abundant for aromatic than for aliphatic species[46]. In general the sensitivity is two or three orders of magnitude lower than for positive ion production and workers in this field are still in the exploratory stage. The use of chemical ionisation sources[4] and atmospheric pressure sources[5] should improve the sensitivity.

1.2.6 Data acquisition and handling

Although for most purposes data acquisition has been standardised, the handling of data to produce chemical results is still the subject of much research and thought. Most effort is going in the direction of retrieval systems whose principal purpose is to find the nearest match to a given spectrum in a library of spectra. Typical of such systems is one described which accepts information on molecular weight and molecular formula, if available, peak mass and intensity information, for a few peaks only[47]. Such methods attempt to simplify the search as much as possible. More complex systems have been devised based on the techniques of the 'learning machine' method described in the last article of this series, but, instead of using a series of matching factors which are purely empirical, tries to introduce some of the logic applied by the chemist in analysing a spectrum[48]. The self-training interpretive and retrieval system (STIRS) uses a number of matching factors, including the following:

MF1: homologous series of ions $\leqslant m/e$ 99.
MF2: three most abundant odd and three most abundant even mass peaks $< m/e$ 90.
MF3: three most abundant odd and three most abundant even mass peaks $91 < m/e < 150$.
MF4: three most abundant odd and three most abundant even mass peaks $m/e > 149$.
MF5: small primary neutral losses.
MF6: large primary neutral losses.
MF10: most abundant odd and even mass peak in each 14 mass unit interval.

A preliminary reduction of the file is carried out using MF10. Following this the composite matching factor MF11 is used:

$$MF11 = (MF1 + MF2 + 2MF3 + 2MF4 + 4MF5 + 2MF6)/12$$

Values of MF11 are high when the unknown compound and the reference compound are the same or when they have structural features in common. The reference compound with the highest value of MF11 is most likely to correspond in structure with the unknown. Where the highest value of MF11 is relatively low, it is likely that there is no compound in the file with structural features comparable with those of the unknown.

Even more complex is the Stanford DENDRAL system, which uses a set of fragmentation rules and generates possible structures. Its use is usually successful when limited to given structural type, and its extension to the structure determination of oestrogenic steroids is as successful as previous applications to monofunctional aliphatic compounds[49]. An automated system has been devised to acquire and reduce metastable data using a slow magnetic scan at high resolution combined with a fast scan of the electrostatic analyser of a double-focussing mass spectrometer[50]. Such a system must be limited to specialised uses, but it does obtain automatically complete information available on fragmentation pathways, which would be extremely tedious to obtain manually. A program has been devised for the calculation of formulae of ions derived from spectra containing polyisotopic atoms[51]. This can handle several polyisotopic atoms in the same molecule and should be useful for the analysis of the spectra of organometallic compounds.

1.2.7   Techniques for investigating ion structure

Among the problems which beset those who attempt to investigate the structure of ions produced in the mass spectrometer, two which often arise involve the randomisation of the atoms in isotopically labelled molecules after ionisation and the production of ions with identical properties from quite diverse neutral antecedents. The latter problem of identity is difficult because different techniques encounter the ions at different times after ion formation, and may only affect a small proportion of the ions of that mass which have a particular internal energy. For example, it has often been assumed that if two ions of the same mass decompose by two competing pathways for both of which metastable ions can be observed, then the intensity ratios of the two metastable ions will be the same if the parent ions have the same structure. The fault in this argument is that the metastable ions are a long-lived group of ions which may have rearranged since they were formed and may not be typical of the ions immediately after formation. They are also ions which decompose and may not be typical of ions which do not decompose.

A more refined method of looking at metastable ions is to examine their peak shapes. It is possible to calculate from these the kinetic energy released in the decomposition[52]. This is usually done by measuring the peak width at half-height. If this is measured accurately and, using a variable β-slit, extrapolated to zero slit-width, it is found that the variation in kinetic energy release with the method of preparation of the ion is observable but not large[53]. An example of this is shown in the study of the loss of HCN from the molecular ions of substituted benzaldehyde oxime methyl ethers. There are two simple mechanisms which can be considered, (12) → (13) and (12) → (14).

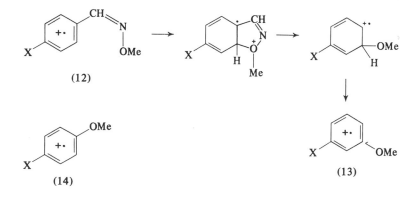

A distinction can be made by analysis of the metastable peak for the process $ArOMe^{+\cdot} \rightarrow ArH^{+\cdot}$. Peak shape analysis shows that this process in (12; X = Cl) has the same kinetic energy release as that for m-chloroanisole and is quite different from those for o- and p-chloroanisole[54]. Reversal of electric and magnetic fields in an instrument of Nier–Johnson geometry allows easier detection, identification and resolution of metastable ions[55].

There is a measurable isotope effect on the kinetic energy release (T) during the loss of H· (or D·) from a metastable molecular ion. For benzene, and a number of other molecules, $T_H/T_D \leqslant 1$, but for Ph_3PO, $T_H/T_D = 1.23$ [56]. It is suggested that this is due to the involvement of H/D in bond formation as well as in bond fission in the suggested mechanism for this process, as in (15) → (16).

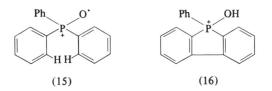

(15)   (16)

In the mass spectrum of dimethyl ether the metastable ion for the process

$$CH_3OCH^{+\cdot} \rightarrow CH_3OCH_2^+ + H\cdot$$

in the first field-free region of a double-focusing mass spectrometer is broadened and shifted in position by an increase in gas pressure in the field-free region. The shift of 3.6 eV on the energy scale corresponds to the amount of kinetic energy converted to internal energy in the collision[57]. Similar shifts measured for the process

$$M^+ + N \rightarrow M^{2+} + N + e^-$$

give good values for $IP^{II} - IP^I$ for a number of molecules[58] and can even be used to study the reaction

$$M^{2+} + N \rightarrow M^{3+} N + e^-$$

The use of a collision gas in the second field-free region of a reversed field double-focusing mass spectrometer allows the measurement of collision-induced metastable spectra (collisional activation spectra, or CA) for focused ions[59]. Such spectra can be used in the same manner as normal metastable peaks to compare the structures of ions. Using this method it appears that there are only two types of stable ion of formula $C_2H_6N^+$, probably $CH_3CH\overset{+}{=}NH_2$ and $CH_2\overset{+}{=}NHCH_3$ [60]. The latter is formed both from molecules with the structural feature CH_3NH-CH_2- and from tetramethyl-hydrazine. The results appear to be more positive than those from normal metastable ions[61]. Collision gases can also be used in double-focusing systems to produce doubly charged ion spectra. In this case there is charge transfer but little momentum transfer to the gas molecule:

$$M^+ + N \rightarrow M^+ + N^+$$

The single charged ions formed by this process have twice the kinetic energy of normal ions and will be observed at twice the electrostatic analyser voltage[61].

Reactivity in bimolecular processes is another probe used for ion structure, often using ICR double resonance techniques. $C_5H_8O^{+\cdot}$ from 2-propylcyclopentanone gives a protonation reaction with ketones at short residence times:

$$C_5H_8O^{+\cdot} + CD_3COCD_3 \rightarrow C_5H_7O\cdot + (CD_3)_2\overset{+}{C}OH$$

This is a reaction typical of an enol ion. At longer residence times the reaction involved is

$$C_5H_8O^{+\cdot} + CD_3COCD_3 \rightarrow C_5H_8O\overset{+}{C}OCD_3 + CD_3 \cdot$$

which is a reaction typical of ketone molecular ions[62]. The enol ion from the McLafferty rearrangement of 2-propylcyclopentanol is ketonising in $ca.$ 10^{-3} s. Both CA spectra and bimolecular reactivity are characteristics of lower energy ions which do not have enough energy to fragment. Photochemical reactions of ions can be observed by ICR. In the photo-dissociation of alkylbenzene molecular ions, the onsets of reaction are greatly in excess of thermochemical requirements for the processes observed, e.g.

$$C_6H_5CH_2R^{+\cdot} \rightarrow C_7H_7^+ + R\cdot$$

There is correspondence, however, with the energy difference between the excited and the ground state as calculated from photoelectron spectra[63]. Photo-dissociation curves are approximate electronic absorption spectra of the ionic species present.

1.3 INVESTIGATION OF ION STRUCTURES

1.3.1 Aromatic species

The behaviour of the molecular ion of benzene is not well understood. The normal electron impact mass spectrum of $[1,2\text{-}^{13}C_2, 3,4,5,6\text{-}^2H_4]$ benzene is consistent with almost complete and independent randomisation of carbon and hydrogen atoms, although it could also be explained by randomisation of hydrogen alone. Metastable ion spectra give more definite evidence that randomisation of all atoms is complete[64]. The metastable ion intensities in spectra of C_6H_6 and C_6D_6 show little change for the $[M - H_2]^+$ and $[M - H]^+$ processes, but a substantial change for the $[M - C_2H_2]^+$ and $[M - CH_3]^+$ processes[65]. This would be consistent with the existence of two isolated states of $C_6H_6^{+\cdot}$ from benzene, one which decomposes to $C_6H_5^+$ and the other to $C_4H_4^+$, a suggestion which has been made on the basis of measurements of ion decomposition kinetics[66]. It is possible to examine the metastable decomposition of $C_6H_6^{+\cdot}$ formed by the following process:

$$C_6H_6 \rightarrow C_6H_6^{2+}$$
$$C_6H_6^{2+} + M \rightarrow M^{+\cdot} + C_6H_6^{+\cdot}$$

The kinetic energy release for the decomposition of this ion is very similar to that observed when the $C_6H_6^{+\cdot}$ is formed by direct ionisation of benzene. It is then argued that since $C_6H_6^{2+}$ is linear, the singly charged ion derived from it will be linear and so the decomposing $C_6H_6^{+\cdot}$ ions in the benzene spectrum must also be linear[67].

There is evidence that non-dissociating $C_7H_8^{+\cdot}$ ions from cycloheptatriene, norbornadiene and toluene retain their identity. Photo-dissociation curves of the three ions are quite different[68], and the following reaction of the toluene molecular ion is not observed with the ions from cycloheptatiiene oi nor-bornadiene[69]:

$$C_7H_8^{+\cdot} + RONO_2 \rightarrow C_7H_8NO_2^+ + RO\cdot$$

The same sort of distinction cannot be made for $C_7H_8^{+\cdot}$ ions which decompose. The kinetic energy release is the same for a number of reactions of the type

$$C_7H_8^{+\cdot} \rightarrow C_7H_7^+ + H\cdot$$

in which $C_7H_8^{+\cdot}$ is formed as a molecular or fragment ion from a variety of sources[70]. The ion $C_7H_8^{+\cdot}$ from labelled $PhCH_2CH_2OH$ shows loss of H· to form $C_7H_7^+$ with randomisation of all hydrogens, both in normal metastable ions and by collisional activation. The corresponding ion which is formed with enough energy to undergo the $C_7H_7^+ \rightarrow C_5H_5^+$ metastable transition has a shorter lifetime, and analysis of these metastables in deuterated species shows that the hydrogen atom lost comes from the original OH or ring CH rather than from the α-CH_2. The conclusion in this case is that the structure of $C_7H_8^{+\cdot}$ immediately after rearrangement is (17) rather than ionised toluene[71].

(17)

Kinetic energy release from the metastable transition

$$C_7H_7^+ \rightarrow C_5H_5^+ + C_2H_2$$

is 30 meV from cycloheptatriene and toluene and 46 meV from benzyl chloride. From similar measurements it is suggested that cycloheptatriene, toluene, the methoxytoluenes and cycloheptatrienyl methyl ether form a tropylium ion, and benzyl chloride, benzyl methyl ether and n-butylbenzene form predominantly a benzyl ion[70]. Results of CA spectra are consistent with this in that the spectrum of $C_7H_7^+$ from 1,2-diphenylethane (benzyl) is different from the common ion from toluene, cycloheptatriene and norbornadiene (tropylium)[72].

In contrast to the behaviour of the related hydrocarbon species, hydrogen scrambling is incomplete in the process

$$C_7H_8Cr^+ \rightarrow C_7H_7Cr^+$$

in the mass spectrum of labelled cycloheptatrienechromium tricarbonyl. The same is true of $C_7H_7^+$ formation from the same molecule[73]. All of the molecules (18)–(21) form of $C_7H_6F^+$ ion which decomposes from a common structure[74].

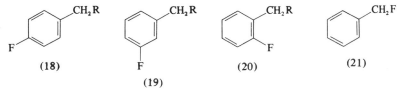

(18)   (19)   (20)   (21)

The $C_9H_{11}^+$ ion from compounds of the general formula $C_6H_5C_3H_6X$ undergo a reaction

$$C_9H_{11}^+ \rightarrow C_7H_7^+ + C_2H_4$$

in which all six hydrogen atoms of the side-chain become randomised prior to fragmentation[75], adding support to a previous suggestion that this ion is a phenylated cyclopropane[76]. Other labelling studies have cast doubt on this interpretation[77]. Metastable ion studies and ^{13}C labelling results suggest that the $C_{11}H_9^+$ ion from (22), (23) and (24) is benztropylium[78]. Thermodynamic measurements confirm this[79].

The presence of certain heteroatoms appears to decrease the uncertainty in interpretation of mass spectra. The elimination of ketene from acetanilide is a case in point. From an analysis of the effect of deuterium labelling on the relative intensities of competing metastable ions in the mass spectra of *p*-chloroaniline and *p*-chloroacetanilide, it was shown that the $[M - C_2H_2O]^+$ ion from the latter compound behaves like (22) rather than (23)[80].

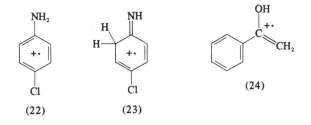

McLafferty rearrangement products (24) from aryl ketones undergo a metastable transition involving loss of $CH_3\cdot$. Such ions do not appear to ketonise since deuterium labelling indicates involvement of ring hydrogen atoms in the $CH_3\cdot$ loss[81].

The technique of ion kinetic energy spectroscopy (IKES) involves the detection of metastable transitions using an electrostatic analyser without mass resolution. This method is very sensitive and is appropriate for the analysis of isomers which will give identical spectra if they rearrange to a common ion, or randomise. The IKES of α- and β-chloronaphthalene are identical, suggesting chlorine randomisation in at least one ring[82]. Di- and tri-chlorobenzene isomers show evidence of incomplete randomisation[83]. Of the isomeric dichlorobiphenyls, four are similar, but (25) and (26) show pronounced differences from the others. The differences are more pronounced between

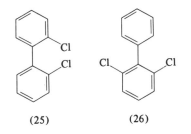

the tetrachlorobiphenyl isomers[83], where chlorine randomisation must be slower on the metastable ion timescale. The *ortho-* and *para-*isomers of DDT, (27) and (28), can be easily distinguished by their IKES, whereas their mass

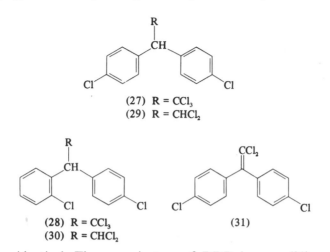

(27) R = CCl₃
(29) R = CHCl₂

(28) R = CCl₃
(30) R = CHCl₂

(31)

spectra are identical. The same is true of DDD isomers (29) and (30)[84]. Isomers of DDE (31) cannot be so identified.

1.3.2 Aliphatic species

INDO calculations suggest that diborane-like structures may be the intermediates in the hydrogen randomisation of the ethane molecular ion. The calculated energy for this species is close to the measured appearance potential of CH_3^+ [85]. $C_2H_5^+$ ions from deuterium labelled ethyl halides and propane undergo proton transfer to NH_3 and hydride transfer from isobutane. The products show complete randomisation of H and D with a rate constant of *ca.* 10^{10} s^{-1} [86]. In an analogous solution reaction, the ionisation of CD_3CH_2F in SbF_5–SO_2 at $-78°C$, hydride transfer from C_6H_{12} gives ethane in which H and D are completely randomised.

Differences of metastable intensities of less than a factor of 5 are not usually considered to be indicative of differences in parent ion structure. Consistent but small differences can be observed in $C_3H_6^+$· ions from a number of sources, suggesting that there are two possible structures which are not completely interconvertible in this timescale[88]. The reactivities of $C_3H_6^+$· from propylene and cyclopropane with ammonia are quite distinct. From propylene only NH_4^+ is formed, but cyclopropane ions also form CH_5N^+ and CH_4N^+ [89]. The relative abundance of NH_4^+ produced from cyclopropane-derived $C_3H_6^+$· ions increases with increasing photon energy. This is interpreted as a result of acyclic ion formation. An ICR study of $C_3H_4D_2^+$ produced from [2,2,5,5-²H₄] tetrahydrofuran shows that it undergoes the same reactions with ammonia as does ionised cyclopropane[90]. Reaction products using ND_3 show that there is little randomisation of hydrogen atoms bonded to carbon and nitrogen atoms. The deuterium content of the CH_4N^+ and

CH_5N^+ ions is consistent with the formation of a cyclic $C_3H_6^+\cdot$ ion in which the carbon atoms have lost their individual identity.

In an ICR study of ethylene the ion $C_3H_5^+$ produced by the reaction

$$C_2H_4^+\cdot + C_2H_4 \rightarrow C_3H_5^+ + CH_3$$

undergoes the photochemical process

$$C_3H_5^+ + C_2H_4 \rightarrow C_3H_7^+ + C_2H_2$$

with an onset of 5200 Å (2.5 eV). This value puts the energy of the first singlet state of $C_3H_5^+$ much lower than calculations suggest[91].

$C_3H_7O^+$ ions from a variety of compounds appear to exist in only four stable forms according to CA spectra. The suggested structures are (32)–(35)[92]. Extrapolating from this result one would expect to observe six different stable $C_4H_9O^+$ ions.

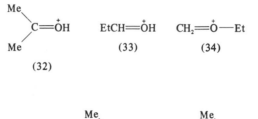

(32)   (33)   (34)

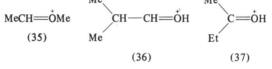

(35)   (36)   (37)

Measurements of abundances of competing metastable ions suggest that only five different forms exist and that compounds which would be expected to form (36) or (37) all form the same $C_4H_9O^+$ ion[93].

Confirming previous evidence from ICR spectra, the CA spectra of the McLafferty rearrangement products $C_3H_6O^+\cdot$ from pentan-2-one, nonan-2-one and nonan-5-one are almost identical and quite distinct from that of the molecular ion of acetone[94]. In the decomposition of n-pentyl vinyl thioether

$$C_5H_{11}SC_2H^+\cdot \rightarrow C_2H_4S^+\cdot + C_5H_{10}$$

the hydrogen atom transferred to the vinyl thio group comes from C-2, C-3 and C-4 of the pentyl group. ICR studies[95] of a variety of sulphur compounds specifically deuterated show that those ions formed by hydrogen transfer from C-2 have the structure $CH_3CH{=}\overset{+}{S}\cdot$, while those involving transfer of C-3 and C-4 hydrogens have the structure $CH_2{=}CH{-}\overset{+\cdot}{S}H$.

1.3.3 Doubly charged ions

Doubly charged ion spectra of hydrocarbons tend to give common ions, and are less sensitive to the structure of the parent neutral than are singly charged spectra. A considerable amount of hydrogen stripping is observed and it is suggested[61] that the frequently observed series of ions $C_nH_2^{2+}$ and $C_nH_6^{2+}$ may be

$$H\overset{+}{-}C(=C=)_n\overset{+}{C}-H \quad \text{and} \quad CH_3-\overset{+}{C}(=C=)_n\overset{+}{C}-CH_3$$

Aromatic sulphur-, oxygen- and nitrogen-containing compounds which exhibit skeletal rearrangements in their singly charged ions also do so in their doubly charged ions[96]. The doubly charged toluene molecular ion shows a greater preference for loss of an α- rather than a ring-hydrogen atom that does the corresponding singly charged ion. The formation of $C_7H_6^{2+}$, however, is preceded or accompanied by complete scrambling of hydrogen atoms[97]. Aromatic amines have much more abundant M^{2+} ions than hydrocarbons, and the principal fragmentation process is loss of C_2H_2 rather than of HCN [98]. Multiply charged ions are formed by field ionisation, and unlike singly charged ions their field-intensity curves show a maximum[99]. Doubly charged acetophenone ions undergo a rearrangement process (38) → (39) which is absent in the singly charged ion spectrum, and requires a strongly electron-releasing group in the *para*-position[100].

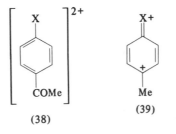

(38)   (39)

Charge separation decomposition processes of doubly charged ions are of interest because they are accompanied by a large release of kinetic energy, which can be attributed to a reduction in electrostatic energy in the process. The measurement of this kinetic energy release allows the calculation of the distance between centres of charge in the decomposing ion. The value for benzene can only be interpreted in terms of an open chain structure for $C_6H_6^{2+}$ and a similar value is found for toluene[101]. The three phenylendiamine isomers all give the same kinetic energy release in charge separation processes and it is suggested that they form a common doubly charged molecular ion[98]. In doubly charged pyrazine, similar measurements indicate that the charges are only four atoms apart and the structure $HC\equiv\overset{+}{N}-CH=CH-\overset{+}{N}\equiv CH$ has been suggested[102]. In the field ion mass spectrum of benzene, metastable ions are observed for the process.

$$C_6H_6^{2+} \rightarrow C_5H_3^+ + CH_3^+$$

and the kinetic energy release is the same as that measured from the electron impact spectrum[103].

1.4 ION–MOLECULE REACTIONS

1.4.1 Gas-phase acid–base relationships

Although all ion–molecule reactions can be considered to be Lewis acid–Lewis base interactions, there are a number of experiments which produce results which are relevant to the understanding of acid–base relationships in solution. The relative basicities of amines can be studied in a number of ways. The equilibrium

$$B_1H^+ + B_2 \rightleftharpoons B_1 + B_2H^+$$

may be directly observed using a high-pressure ion source and measuring both ion currents and the partial pressures of both species. Where there is a considerable difference in basicity it may be possible to observe by ion cyclotron double resonance that the reaction proceeds in only one direction. The third method which can be applied uses pulsed ion cyclotron resonance, and by following the changes in B_1H^+ and B_2H^+ ion currents the rates of forward and reverse reactions can be measured. The simplest quantitative value which can be used for the basicity of an amine (or other base) is the proton affinity (PA) which is defined as the negative of the enthalpy of the reaction

$$B + H^+ \rightarrow BH^+$$

The first and third methods mentioned above are more accurate than the second, where limits for the proton affinity of a base can be obtained by 'bracketing' it between a stronger base and a weaker base both of known PA.

The most complete set of PAs of amines has been obtained by Yamdagni and Kebarle by the high-pressure technique, and is shown in Table 1.1[104]. A

Table 1.1 Proton affinities/kcal mol⁻¹

NH_3	207	Cyclohexylamine	227
CH_3CONH_2	210	$C_6H_5N(CH_3)_2$	229
Pyrrole	214	$(CH_3)_3N$	230
$C_6H_5NH_2$	216	$C_6H_5N(CH_3)C_2H_5$	231
CH_3NH_2	218	Pyrrolidine	231
$C_6H_5NHCH_3$	222	$C_6H_5N(C_2H_5)_2$	234
$o\text{-}CH_3C_6H_4NH_2$	222	$H_2NCH_2CH_2NH_2$	235
$(CH_3)_2NH$	225	$H_2N(CH_2)_3NH_2$	243
$C_6H_5NHC_2H_5$	225	$H_2N(CH_2)_5NH_2$	243
Pyridine	226	$H_2N(CH_2)_7NH_2$	243

simpler method of analysis of basicity is the use of ammonia, CH_3NH_2, $(CH_3)_2NH$ and $(CH_3)_3N$ as reactant gases in chemical ionisation[105]. The results allow 'bracketing' of bases between the values shown in Table 1.1 for these gases and the order of basicity found in Table 1.2 is consistent with values in Table 1.1 but gives less thermochemical information.

Table 1.2   Order of basicities of amines

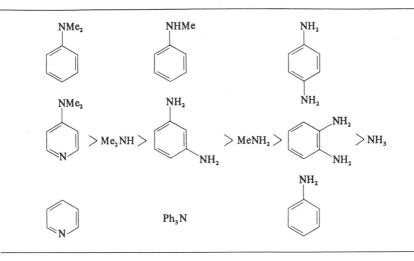

ICR results on aliphatic amines show that both in the series (40) and (41)

$$NH_3, CH_3NH_2, (CH_3)_2NH, CH_3N \qquad (40)$$

$$CH_3NH_2, CH_3CH_2NH_2, (CH_3)_2CHNH_2, (CH_3)_3CNH_2 \qquad (41)$$

there is a significant increase in PA with methyl substitution, but it is greater for (40) than for (41)[106]. The order of basicities in solution

$$NH_3 < primary < secondary > tertiary$$

is different from the gas-phase order. There is no simple explanation since most of the contributing thermodynamic properties change in an orderly way with alkyl substitution, but there are differences in the rate of change of these properties in response to substitution[107]. A plot of gaseous PA vs. heat of protonation in solution gives three separate straight lines for primary, secondary and tertiary amines. The difference here is in the number of $\overset{+}{N}$—M⋯OH$_2$ bonds made in the solvated ion[108]. As shown in Table 1.1, the aliphatic diamines appear to have abnormally high proton affinities. A detailed study of the difference in PA between $CH_3(CH_2)_nNH_2$ and $H_2N(CH_2)_nNH_2$ showed that this was a maximum for $n = 4$ and that a cyclic hydrogen-bonded form such as (42) must be responsible for the high basicity[109].

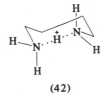

(42)

A bracketing technique has been used to measure proton affinities of a variety of organic oxygen compounds from proton exchange reactions in a pulsed time-of-flight mass spectrometer with delay times of up to 7 μs. The additive effects of longer alkyl chains tend to a constant value where the chain

Table 1.3

R > C$_3$H$_7$	PA/kcal mol^{-1}
RCH$_2$OH	189
R$_2$CHOH	197
R$_3$COH	204
HCO$_2$R	198
CH$_3$CO$_2$R	207
RCO$_2$CH$_3$	205
RCO$_2$R	210

is greater than three atoms long, and the following limits are given (Table 1.3)[110]. Diethers show the same tencency as diamines to form cyclic protonated ions[111]. Both proton affinities and methyl cation affinities (MCA) have been measured for a number of compounds[112]:

$$PA: \quad H_2O > CH_3F > HI > CO > HCl > HF > N_2$$
$$MCA: \quad CO > HI > H_2O > HCl > CH_3F > N_2 > HF$$

The methyl cation affinity is a measure of basicity towards a soft acid while PA is basicity towards a hard acid.

The measurement of equilibria for proton transfer between anions allows the determination of gas phase acidities. For binary hydrides these are as follows:

$$CH_4 < NH_3 < H_2O < HF$$
$$\wedge \qquad \wedge \qquad \wedge \qquad \wedge$$
$$SiH_4 < PH_3 < H_2S < HCl$$
$$\wedge \qquad \qquad \wedge$$
$$AsH_3 \qquad < \qquad HBr$$
$$\wedge$$
$$HI$$

The trend along the rows is a result of electronegativities whereas the trend down the columns reflects the strength of H—X bonds[113]. Acidities have also been measured for phenols[114] and carboxylic acids[115]. Acidities of hydrocarbons are difficult to measure because of low kinetic acidities. The reactions

$$CH_3OH + CH_2{=}CHCH_2^- \rightarrow CH_3O^- + CH_3CH{=}CH_2$$

and

$$C_6H_5CH_3 + CH_3O^- \rightarrow CH_3OH + C_6H_5CH_2^-$$

can be observed but proton transfer between the two carbanions is too slow to be observed[116]:

$$C_6H_5CH_3 + CH_2{=}CH{-}CH_2^- \rightarrow C_6H_5CH_2^- + CH_2{=}CH{-}CH_3$$

Relative carbonium ion stabilities cannot be directly observed using hydride transfer reactions because these are often too slow. Halide transfer reactions are easier to measure. Fluoride affinities of fluoromethyl cations follow the order

$$CF_3^+ > CH_3^+ > CHF_2^+ \approx CH_2F^+$$

while a different order is found for hydride affinities derived from these:

$$CH_3^+ > CF_3^+ > CH_2F^+ > CHF_2^+$$

The hydride affinities are a better measure of carbonium ion stability because C—F bond energies in fluorinated methanes vary considerably more than do C—H bond energies with changes in fluorine substitution[117]. Similar measurements of fluorochloroalkyl carbonium ions have been made[118].

1.4.2 Gas-phase solvation of ions

The equilibrium constant $K_{n,n+1}$ can be measured for the association process where X may be a cation or anion and M a solvent molecule:

$$X \cdot M_n^\pm + M \rightleftharpoons X \cdot M_{n+1}^\pm$$

The value of n which can be observed depends on the temperature and pressure in the ion source. There is still disagreement about the value of $K_{1,2}$ for the $H(H_2O)_n^+$ system, where values as widely apart as 3×10^8 and 1.8×10^{18} have been found by different groups[119,120].

The solvation of halide ions by hydrogen-bond donors has been investigated. The strength of the bond increases with the gas-phase acidity of the donor[121]. If the same 'solvent' is used with different anions, the strength of the 0,1 association increases with the gas phase basicity of the anion. The situation is more complex with larger clusters. In a comparison of water, methanol and dimethyl ether as solvents for the proton, it is found that there is in all cases a decrease in association energy for the $n,n+1$ equilibrium as n increases[122]. This decrease is greater for methanol than for water. For dimethyl ether, the strongest base, there is a drastic fall-off:

$$\Delta G_{2,1} = 21.9 \text{ kcal mol}^{-1}$$
$$\Delta G_{3,2} = 1.9 \text{ kcal mol}^{-1}$$

1.4.3 Other ion–molecule reactions

Acid-catalysed dehydration processes have been observed in a number of alcohols by ICR, e.g. in butan-2-ol[123]:

$$CH_3\overset{+}{C}HOH + C_4H_9OH \rightarrow CH_3\text{—}\overset{+}{C}HOH \cdots OH_2 + C_4H_8$$

Elimination processes initiated by anions are much more closely analogous to base-catalysed eliminations in solution, and are observed by ICR in mixtures of perdeuteromethanol and fluoroethanes[124]:

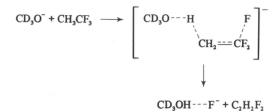

$$CD_3OH---F^- + C_2H_2F_2$$

A gas-phase analogy for acid-catalysed esterification has also been found, in the reaction:

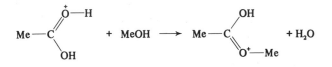

Such a reaction is not found with protonated formic acid since the proton affinity of formic acid is less than that of methanol and proton transfer predominates[125].

Gas-phase aromatic substitution can be followed by examining the addition of charged electrophiles to substituted benzene. Acetylation with $(CH_3CO)_2^{+\cdot}$ follows normal substituent effects but nitration with $CH_2ONO_2^+$ gives an inverse substituent effect[126]. An aromatic methylation reaction has been observed, in which one of the benzene hydrogens is removed[127]:

$$CH_3OCH_2^+ + C_6H_6 \rightarrow C_7H_7^+ + CH_3OH$$

This shows normal substituent effects on reaction rates, but no positional selectivity in the reaction with o, m- or p-[$^2H_1$]toluene.

1.5   DETERMINATION OF MOLECULAR STRUCTURE

1.5.1   Amino acids and peptides

Chemical ionisation spectra have been reported for simple amino acids using methane and isobutane as reagents. Comparisons made at various source temperatures suggest that, using methane, the ion temperatures are ca. 160°C in excess of the source temperature[128]. In a field-desorption study, $M^{+\cdot}$ or $[M + H]^+$ ions were observed for all amino acids, including cystine and arginine, which present problems with CI and EI[129].

Electron impact mass spectra have been used to identify thiohydantoins produced in the Edman procedure. Methyl-, phenyl- and bromophenyl-thiohydantoins can be used and leucine can be distinguished from isoleucine either by fragment ion intensities[130] or metastable ions[131]. Low electron energy results in an increase in the relative abundance of the molecular ion, but an absolute loss in sensitivity[132]. The phenylthiourea derivatives of amino acids are converted into phenylthiohydantoin derivates by thermal reactions in the mass spectrometer inlet. The same sort of reaction (43) → (44) is found with the phenylthiourea derivatives of peptides[133]. This provides a simplified method of determination of the N-terminal.

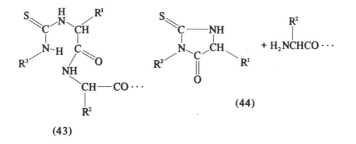

(43)   (44)

In most applications, volatile derivatives of peptides are used and a considerable amount of work has gone into producing methods of N-methylation which avoid sulphonium ion formation from methionine, C-methylation of Gly, Asp and Glu and quaternisation of histidine. The use of $NaH–CH_3I–HCON(CH_3)_2$ [134], $NaH–Me_4SO_4–MeCN$ [134] and $NaH–CH_3I–(CH_3)_2SO$ [135,136] in carefully controlled ratios with respect to the peptide achieves these aims. The arginine (45) side-chain presents difficult problems and requires a specific solution: reaction with a β-diketone to form a pyridine derivative of ornithine (46). Using this reaction and standard derivative

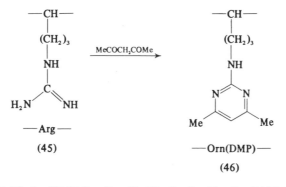

(45)   —Orn(DMP)—
                                                                        (46)

CF_3CO-Orn(DNP)-Pro-Pro-Gly-Phe-Ser-Pro-Phe-Orn(DNP)-OMe
(47)

techniques the permethylated derivative of (47) is produced from bradykinin. In its mass spectrum, peptide sequence peaks can be observed as far as the eighth residue from the N-terminal end[137].

It has been suggested that Schiff base derivatives of peptides may be useful in sequencing studies[138]. Their spectra have abundant molecular ions (the abundance depending upon the aldehyde used) and yield C-terminal as well as N-terminal fragments. Internal fragments are also formed which can be interpreted and used for sequence information. Permethylated acyl peptide esters need not be pure in order to afford sequence information[139]. Identification of the sequences of constituents of peptide mixtures is possible, e.g.

Leu-Leu-Gly-Asn-Val-Leu-Phe in the presence of Leu-Leu-Val-Val-Tyr-Pro-Trp[140]. The spectra of mixtures of derivatised peptides can be analysed using a computer program which can cope with metastable information, chemical ionisation spectra and a series of spectra produced by fractional evaporation of the sample[141].

Field desorption spectra of peptide derivatives have abundant $[M + H]^+$ ions for compounds which do not have a stable M^+ ion in the EI spectrum, e.g. Ac-Gly-Arg-Arg-Gly-OMe. Unambiguous molecular weight information is also possible from free peptides, e.g. Phe-Asp-Ala-Ser-Val. Sequence information in the spectra is often incomplete, however, and EI or CI spectra of derivatives may be necessary for complete structure determination[142]. Photoionisation spectra using a hydrogen lamp (α-line, 10.1 eV) of peptides are similar to EI spectra but have a greater abundance of heavier ions and a lower abundance of lighter ions[35]. As a result, it may be difficult to distinguish Gly-Leu- from Ala-Val- at the N-terminal when the lower mass sequence peaks are absent. Possibly the presence in the hydrogen of an impurity with a higher energy photon emission would help to solve this problem without running two spectra on the same sample.

Biemann has used a unique method of chemical treatment of oligopeptides for mass spectrometric sequencing. The chemical sequence used is acetylation, lithium aluminium deuteride reduction and trimethylsilylation to give compounds of general formula $MeCD_2[NHCHRCD_2]_nOTMS$. These can be sequenced easily by identification of amine fragments. This method has been used successfully with the peptide (48), which was hydrolysed to a mixture of 27 di-, tri- and tetra-peptides.

Trp-Ile-Thr-Lys-Gln-Glu-Tyr-Asp-Glu-Ala-Gly-Pro-Ser-Ile-Val-His-Arg-Lys-AEtCys-Phe
AEtCys = aminoethylcysteine

(48)

Separation and sequencing of these small peptides allowed the complete sequence to be deduced, although a separate experiment was necessary to determine the position of glutamine, which forms glutamic acid under the conditions used[143]. The structure of α-MSH (49) was determined by more conventional methods[144]. The protein was broken down into four fragments, using CNBr, and enzymic digestion, and the fragments analysed as permethylated acyl peptide esters.

Ac-Ser-Tyr-Ser-Met-Glu-His-Phe-Arg-Trp-Gly-Lys-Pro-Val-NH₂

(49)

Peptides containing β-lysine can be distinguished from those containing α-lysine[145], principally by the presence of m/e 70 (50) rather than m/e 84

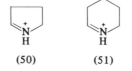

(50)   (51)

(51)[146]. The other unique feature of acylated β-lysine peptides is the presence of ions of the general structure

$$\begin{array}{c} (CH_2)_3NHCOMe \\ | \\ H_3\overset{+}{N}-CH-CH_2COR \end{array}$$

N-Terminal pyroglutamyl peptides related to thyrotropin releasing hormone (TRH) (52) usually give molecular ions and fragments related to the pyroglutamyl residue and the *C*-terminal residue but not the normal sequence fragments typical of acyl peptide esters[147]. However, the fragments are sufficient to allow a sequence determination. The spectrum of (52) has been

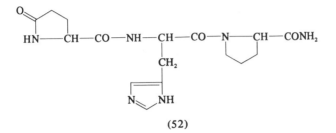

(52)

obtained by reaction with $MeNH_3^+$ ions in the collision chamber of a tandem mass spectrometer[148]. The best spectra are obtained when the chamber is lined with Teflon and the sample is evaporated from a Teflon surface. The energy necessary for volatilisation from Teflon is less than that necessary when glass, copper or carbon surfaces are used.

One of the components of Staphylomycin S has been assigned the structure (53), largely by mass spectrometry. Usually depsipeptides containing a

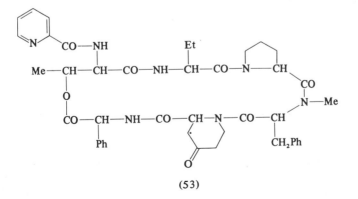

(53)

number of hydroxy-acid residues open by eliminating CO_2. In this case the initial fragmentation was by elimination of dihydropyrid-4-one[149].

1.5.2 Other natural products

It is necessary for the reviewer to be so selective in this field that the choice of included work may appear somewhat arbitrary. The following are some examples of attempts to solve difficult problems or what seem successful solutions to possibly more general problems.

Fructose containing disaccharides and oligosaccharides can be distinguished by mass spectra of suitable derivatives. In the spectra of the trimethylsilyl ethers the $[M - CH_2OTMS]^+$ ion is abundant in the high mass region, while in isomers containing only aldose units the principal ion in this region is $[M - Me - HOTMS]^+$. Another characteristic ion is m/e 437, for which the process (54) → (55) has been suggested[150]. Similar processes are observed in

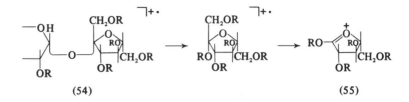

(54)   (55)

permethylated derivatives, but in this case the presence of abundant ions at m/e 88 and 101 are said to be more characteristic[151]. Aldohexose trimethylsilyl ethers can be identified by peak intensities in their spectra[152]. A set of eight intensity ratios R is used. For each isomer a weighting factor a is calculated from the equation

$$S_i = \left[\sum_{n=1}^{8} a_{n_i} R_n \right] + C_i$$

where C is an isomer constant, When an unknown sample is examined, eight values of S are calculated from this equation, one for each isomer. That which gives the highest value of S corresponds to the correct isomer.

Glucose 6-phosphate and the deoxyfluoroglucose 6-phosphates have been examined as their disodium salts by FD mass spectrometry. Abundant $[M + H]^+$ ions are observed. The other principal ions at high mass correspond to Na/H exchange in the formation of the $[M + H]^+$ ion. At lower mass are some inorganic ions and a few organic fragments such as $C_2H_5O_2^+$ and $C_2H_4FO^+$ [153]. Field desorption spectra show a pronounced difference in intensities between the α- and β-isomers of aryl glucosides. All give abundant $M^{+\cdot}$ ions. Such a distinction could be made with difficulty in FI spectra and not at all with EI.

Chemical ionisation affords a method of distinguishing between the isomeric nucleosides (56) and (57)[155]. The ratios of the ions $[M + H]^+$ and $[BH_2]^+$ varied between 4 and 100 and the isomer of structure (56) always has the more abundant $[M + H]^+$. The 2,2′-anhydronucleosides such as (58) can be easily identified by electron impact mass spectra[156]. Field desorption gives good mass spectra of nucleosides and the ions $[M + H]^+$, M^+, $[B + 2H]^+$ and S^+ can all be identified[157]. This is useful in the case of guanosine, which

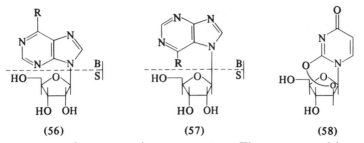

(56)   (57)   (58)

decomposes near the evaporation temperature. The most promising application of FD in this area of chemistry is to nucleotides. The spectrum of 5'-adenosine monophosphate (59) has ions corresponding to $[M + H]^+$,

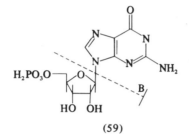

(59)

$[M - PO_3H_2]^+$, $[B + H_2]^+$ and $[B + H]^+$. These allow the three constituent units of the nucleotide to be identified[157].

In the CI spectra of macrolide antibiotics the principal ions observed are $[M + H]^+$ and fragments due to loss of H_2O or cleavage at glucoside bonds[158]. The only C—C bond cleavages observed can be explained as retro-aldol processes, e.g. (60) → (61).

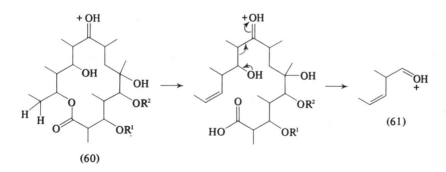

(60)   (61)

1.5.3 Stereochemical problems

A number of examples quoted in the last few pages show that there can be some stereochemical discrimination in fragmentation processes. The spectra of *cis*- and *trans*-t-butylcyclohexanols have been examined by photoionisation with the Lyman α-line[34]. In addition to a gross difference in the $[M^+]/[M^+ - H_2O]$ ratios there is an ion at m/e 110 which is present in the *trans* isomer and absent in the *cis* isomer. The explanation given is that the preferred 1,4-elimination is possible in the *trans* isomer and the bicyclohexane ion (62) can decompose by elimination of ethylene. The product of 1,3-elimination from the *cis* isomer will have a different structure.

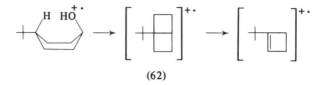

(62)

The specificity of the water elimination process has been used in the determination of the relative stereochemistry of the isomeric alcohols (63) and (64)[159]. Addition of D_2 by homogeneous hydrogenation produces the saturated labelled alcohols (65) and (66). Elimination of HDO is possible only in

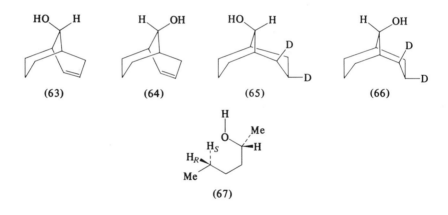

(63)   (64)   (65)   (66)

(67)

the isomer (66). A small but measureable degree of stereospecificity is observed for this process in hexan-2-ol (67)[160]. The rate ratio k_R/k_S for elimination of H_2O from the two diastereotopic positions is 1.16. This is very similar to the value for 1,4-hydrogen abstraction in an intramolecular free radical process.

Compounds of the general structure (68) undergo a retro-Diels–Alder

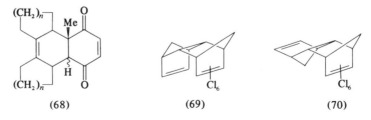

(68)   (69)   (70)

reaction much more easily when the junction is *cis* than when it is *trans*. Rate factors of between 10 and 300 are observed and it is suggested that the process must be concerted[161]. In the chemical ionisation spectra of isodrin (69) and aldrin (70) the retro-Diels–Alder process $[M - Cl]^+ \rightarrow C_7H_2Cl_5^+$ occurs much more easily in (69) than in (70)[162]. In *cis* and *trans* fused isomers of (71), (72) and (73) the effects are unpredictable[163]. *Erythro*- and *threo*-isomers of 2,3-diphenylbutane exhibit small but reproducible differences in

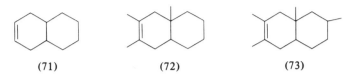

(71)   (72)   (73)

the intensities of peaks in their spectra. These are ascribed to differences in energy between the parent ions rather than to energy differences in transition states[164].

The mass spectra of *cis*- and *trans*-2-substituted cyclopentyl bromides have been recorded[165]. The results are in agreement with the operation of two different effects on the rate of loss of Br·: ground state energy differences and transition state energy differences due to neighbouring group participation. The latter effect is strong for Br, weak for Cl, and almost non-existent for OH and OMe. A chemical ionisation study of compounds of the type X—CH₂— CH₂—OH found somewhat different substituent effects[166]. The order of effectiveness in assisting the process $[M + H]^+ \rightarrow [MH - H_2O]^+$ is Br > SH > SMe > OMe $\geqslant$ Cl, F.

Intramolecular interactions in mass spectra can take place over a much longer distance than in solution chemistry. The elimination of RCOCH₂· from the molecular ion of RCO(CH₂)ₙNH₂ has the highest probability[167] when $n = 4$–8. The process

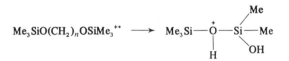

is still observed when $n = 46$. In the absence of a solvent, ion–dipole interactions take place at long range[168].

References

1. Arsenault, G. P., Dolhun, J. J., and Biemann, K. (1970). *Chem. Commun.*, 1542; Schoengold, D. M. and Munson, B. (1970). *Anal. Chem.*, **42**, 1811
2. Baldwin, M. A. and McLafferty, F. W. (1973). *Org. Mass Specrom.*, **7**, 1111
3. Baldwin, M. A. and McLafferty, F. W. (1973). *Org. Mass. Spectrom.*, **7**, 1353
4. Dougherty, R. C., Dalton, J. and Biros, F. J. (1972). *Org. Mass Spectrom.*, **6**, 1171
5. Horning, E. C., Horning, M. G., Carroll, D. I. and Dzidic, I. (1973). *Anal. Chem.*, **45**, 936
6. Bennett, S. L. and Field, F. H. (1972). *J. Amer. Chem. Soc.*, **94**, 8669

7. Clow, R. P. and Futrell, J. H. (1972). *J. Amer. Chem. Soc.*, **94**, 3748
8. Bonner, R. F., Lawson, G. and Todd, J. F. J. (1972). *J. Chem. Soc. Chem. Commun.*, 1179
9. Hunt, D. F. and McEwen, C. N. (1973). *Org. Mass Spectrom.*, **7**, 441
10. Foltz, R. L., Fentiman, A. F., Jr., Mitscher, L. A. and Showalter, H. D. H. (1973). *J. Chem. Soc. Chem. Commun.*, 872
11. Michnowicz, J. and Munson, B. (1972). *Org. Mass Spectrom.*, **6**, 283
12. Arsenault, G. P. (1972). *J. Amer. Chem. Soc.*, **94**, 8241
13. Dzidic, I. and McCloskey, J. A. (1972). *Org. Mass Spectrom.*, **6**, 939
14. Hunt, D. F. and Ryan, J. F. (1972). *J. Chem. Soc. Chem. Commun.*, 620
15. Einoff, B. and Munson, B. (1972). *Int. J. Mass Spectrom., Ion Phys.*, **9**, 141
16. Einolf, N. and Munson, B. (1973). *Org. Mass Spectrom.*, **7**, 155
17. Beckey, H. D., Bloching, S., Migahed, M. D., Ochterbeck, E. and Schulten, H.-R. (1972). *Int. J. Mass Spectrom. Ion Phys.*, **8**, 169
18. Schulten, H.-R. and Beckey, H. D. (1972). *Org. Mass Spectrom.*, **6**, 885
19. Winkler, H. U. and Beckey, H. D. (1973). *Org. Mass Spectrom.*, **7**, 1007
20. Schulten, H.-R. and Beckey, H. D. (1973). *J. Chromatog.*, **83**, 315
21. Tou, J. C. (1972). *Org. Mass Spectrom.*, **6**, 833
22. Beckey, H. D. and Migahed, M. D. (1972). *Org. Mass Spectrom.*, **6**, 923
23. Levsen, K. and Beckey, H. D. (1972). *Int. J. Mass Spectrom. Ion Phys.*, **9**, 51
24. Levsen, K. and Beckey, H. D. (1972). *Int. J. Mass Spectrom. Ion Phys.*, **9**, 63
25. Derrick, P. J., Falick, A. M. and Burlingame, A. L. (1972). *J. Amer. Chem. Soc.*, **94**, 6794
26. Derrick, P. J., Falick, A. M. and Burlingame, A. L. (1974). *J. Amer. Chem. Soc.*, **96**, 616
27. Brown, P. and Fenselau, C. (1974). *Org. Mass Spectrom.*, **7**, 305
28. Derrick, P. J., Falick, A. M. and Burlingame, A. L. (1973). *Org. Mass Spectrom.*, **7**, 890
29. Derrick, P. J., Falick, A. M. and Burlingame, A. L. (1973). *J. Amer. Chem. Soc.*, **95**, 437
30. Atkinson, R., Finalyson, B. J. and Pitts, J. N., Jr. (1973). *J. Amer. Chem. Soc.*, **95**, 7592
31. Jones, I. T. N. and Bayes, K. D. (1972). *J. Amer. Chem. Soc.*, **94**, 6869
32. Stebbings, W. L. and Taylor, J. W. (1972). *Int. J. Mass Spectrom. Ion Phys.*, **9**, 471
33. Johnson, B. M. and Taylor, J. W. (1972/3). *Int. J. Mass Spectrom. Ion Phys.*, **10**, 1
34. Akhtar, Z. M., Brion, C. E. and Hall, L. D. (1973). *Org. Mass Spectrom.*, **7**, 647
35. Orlov, V. M., Varshavsky, Ya. M. and Kiryushkin, A. A. (1972). *Org. Mass Spectrom.* **6**, 9
36. Eland, J. H. D. (1972). *Int. J. Mass Spectrom. Ion Phys.*, **8**, 143
37. Danby, C. J. and Eland, J. H. D. (1972). *Int. J. Mass Spectrom. Ion Phys.*, **8**, 153
38. Simm, I. G., Danby, C. J. and Eland, J. H. D. (1973). *J. Chem. Soc. Chem. Commun.*, 832
39. Mumma, R. O. and Vastola, F. J. (1970). *Org. Mass Spectrom.*, **3**, 101
40. Mumma, R. O. and Vastola, F. J. (1972). *Org. Mass Spectrom.*, **6**, 1373
41. McAllister, T. (1972). *J. Chem. Soc. Chem. Commun.*, 245
42. Yinon, J. and Boetger, H. G. (1972/73). *Int. J. Mass Spectrom. Ion Phys.*, **10**, 161
43. Todd, J. F. J., Turner, R. B., Webb, B. C. and Wells, C. H. J. (1973). *J. Chem. Soc. Perkin Trans. II*, 1167
44. Alexander, R. G., Bigley, D. B. and Todd, J. F. J. (1972). *J. Chem. Soc. Chem. Commun.*, 553; (1973). *Org. Mass Spectrom.*, **7**, 963
45. Bowie, J. H. and White, P. Y. (1972). *Org. Mass Spectrom.*, **6**, 75
46. Ito, A., Matsumoto, K. and Takeuchi, T. (1973). *Org. Mass Spectrom.*, **7**, 1280
47. Heller, S. R., Fales, H. M. and Milne, G. W. A. (1973). *Org. Mass Spectrom.*, **7**, 107
48. Kwok, K.-S., Venkataraghavan, R. and McLafferty, F. W. (1973). *J. Amer. Chem. Soc.*, **95**, 4185
49. Smith, D. H., Buchanan, B. G., Engelmore, R. S., Duffield, A. M., Feigenbaum, E. A., Lederberg, J. and Djerassi, C. (1972). *J. Amer. Chem. Soc.*, **94**, 5962
50. Coutant, J. E. and McLafferty, F. W. (1972). *Int. J. Mass Spectrom. Ion Phys.*, **8**, 323

51. Crawford, L. R. (1972/73). *Int. J. Mass Spectrom. Ion Phys.*, **10**, 279
52. Beynon, J. H., Fontaine, A. E. and Lester, G. R. (1972). *Int. J. Mass Spectrom. Ion Phys.*, **8**, 341
53. Jones, E. G., Bauman, L. E., Beynon, J. H. and Cooks, R. G. (1973). *Org. Mass Spectrom.*, **7**, 185
54. Beynon, J. H., Bertrand, M. and Cooks, R. G. (1973). *Org. Mass Spectrom. Ion Phys.*, **7**, 785
55. Wachs, T., Bente, P. F., III, and McLafferty, F. W. (1972). *Int. J. Mass Spectrom. Ion Phys.*, **9**, 333
56. Bertrand, M., Beynon, J. H. and Cooks, R. G. (1973). *Org. Mass Spectrom.*, **7**, 193
57. Beynon, J. H., Bertrand, M., Jones, E. G. and Cooks, R. G. (1972). *J. Chem. Soc. Chem. Commun.*, 341
58. Cooks, R. G., Ast, T. and Beynon, J. H. *Int. J. Mass Spectrom. Ion Phys.*, **11**, 490
59. McLafferty, F. W., Bente, P. F. III, Kornfield, R. and Tsai, S.-C. (1973). *J. Amer. Chem. Soc.*, **95**, 2120
60. Levsen, K. and McLafferty, F. W. (1974). *J. Amer. Chem. Soc.*, **96**, 139
61. Ast, T., Beynon, H. J. and Cooks, R. G. (1972). *Org. Mass Spectrom.*, **6**, 749
62. Hass, J. R., Bursey, M. M., Kingston, D. G. I. and Tannenbaum, H. P. (1972). *J. Amer. Chem. Soc.*, **94**, 5095
63. Dunbar, R. C. (1973). *J. Amer. Chem. Soc.*, **95**, 6191
64. Beynon, J. H., Caprioli, R. M., Perry, W. O. and Baitinger, W. E. (1972). *J. Amer. Chem. Soc.*, **94**, 6828
65. Howe, I., Uccella, N. and Williams, D. H. (1973). *J. Chem. Soc. Perkin Trans. II*, 76
66. Andlauer, B. and Ottinger, Ch. (1971). *J. Chem. Phys.*, **55**, 1471
67. Keough, T., Ast, T., Beynon, J. H. and Cooks, R. G. (1973). *Org. Mass Spectrom.*, **7**, 345
68. Dunbar, R. C. and Fu, E. W. (1973). *J. Amer. Chem. Soc.*, **95**, 2718
69. Hofmann, M. K. and Bursey, M. M. (1971). *Tetrahedron Lett.*, 2539
70. Cooks, R. G., Beynon, J. H., Bertrand, N. and Hofmann, M. K. (1973). *Org. Mass Spectrom.*, **7**, 1303
71. Levsen, K., McLafferty, F. W. and Jerina, D. M. (1973). *J. Amer. Chem. Soc.*, **95**, 6332
72. Winkler, J. and McLafferty, F. W. (1973). *J. Amer. Chem. Soc.*, **95**, 7533
73. Tibbetts, F. E. III, Bursey, M. M., Little, W. F., Willeford, B. R., Benezra, S. A., Hofmann, M. K. and Jennings, P. W. (1972). *Org. Mass Spectrom.*, **6**, 475
74. Jennings, K. R. and Whiting, A. (1972). *Org. Mass Spectrom.*, **6**, 917
75. Uccella, N. A. and Williams, D. H. (1972). *J. Amer. Chem. Soc.*, **94**, 8778
76. Rylander, P. N. and Meyerson, S. (1956). *J. Amer. Chem. Soc.*, **78**, 5399; Meyerson, S. and Hart, H. (1963). *J. Amer. Chem. Soc.*, **85**, 2385
77. Köppel, C., Schwarz, H. and Bohlmann, F. (1973). *Org. Mass Spectrom.*, **7**, 869
78. Schwarz, H. and Bohlmann, F. (1973). *Org. Mass Spectrom.*, **7**, 23
79. Schwarz, H. and Bohlmann, F. (1973). *Org. Mass Spectrom.*, **7**, 395
80. Uccella, N., Howe, I. and Williams, D. H. (1972). *Org. Mass Spectrom.*, **6**, 229
81. Tomer, K. B. and Djerassi, C. (1972). *Org. Mass Spectrom.*, **6**, 1285
82. Safe, S., Hutzinger, O. and Cook, M. (1972). *J. Chem. Soc., Chem. Commun.*, 260
83. Safe, S., Hutzinger, O. and Jamieson, W. D. (1973). *Org. Mass Spectrom.*, **7**, 169
84. Safe, S., Hutzinger, O., Jamieson, W. D. and Cook, M. (1973). *Org. Mass Spectrom.*, **7**, 217
85. Parker, C. E., Bursey, M. M. and Pederson, L. G. (1973). *Org. Mass Spectrom.*, **7**, 1077
86. Ausloos, P., Rebbert, R. E., Sieck, L. W. and Tiernan, T. O. (1972). *J. Amer. Chem. Soc.*, **94**, 8939
87. Vorachek, J. H., Meisels, G. G., Geanangel, R. A. and Emmel, R. H. (1973). *J. Amer. Chem. Soc.*, **95**, 4078
88. Gross, M. L. and Liu, P. H. (1974). *Org. Mass Spectrom.*, **7**, 795
89. Sieck, L. W., Gordon, R., Jr., and Ausloos, P. (1972). *J. Amer. Chem. Soc.*, **94**, 7157
90. Gross, M. L. (1972). *J. Amer. Chem. Soc.*, **94**, 3744
91. Kramer, J. M. and Dunbar, R. C. (1972). *J. Amer. Chem. Soc.*, **94**, 4347
92. McLafferty, F. W. and Sakai, I. (1973). *Org. Mass Spectrom.*, **7**, 971
93. Mead, T. J. and Williams, D. H. (1972). *J. Chem. Soc. Perkin Trans. II*, 876

94. McLafferty, F. W., Kornfeld, R., Haddon, W. F., Levsen, K., Sakai, I., Bente, P. F. III, Tsai, S.-C. and Schuddemage, H. D. R. (1973). *J. Amer. Chem. Soc.*, **95**, 3886
95. Tomer, K. B. and Djerassi, C. (1973). *J. Amer. Chem. Soc.*, **95**, 5335
96. Blumenthal, T. and Bowie, J. H. (1972). *Org. Mass Spectrom.*, **6**, 1053
97. Ast, T., Beynon, J. H. and Cooks, R. G. (1972). *J. Amer. Chem. Soc.*, **94**, 1834
98. Ast, T. and Beynon, J. H. (1973). *Org. Mass Spectrom.*, **7**, 503
99. Goldenfeld, I. V., Korostyshevsky, I. Z. and Nazarenko, V. A. (1973). *Int. J. Mass Spectrom. Ion Phys.*, **11**, 9
100. Sakurai, H., Tatematsu, A. and Nakata, H. (1973). *Org. Mass Spectrom.*, **7**, 1109
101. Ast, T., Beynon, J. H. and Cooks, R. G. (1972). *Org. Mass Spectrom.*, **6**, 101
102. Beynon, J. H., Caprioli, R. M. and Ast, T. (1972). *Org. Mass Spectrom.*, **6**, 102
103. Beckey, H. D., Migahed, M. D. and Rollgen, F. W. (1972/73). *Int. J. Mass Spectrom. Ion Phys.*, **10**, 471
104. Yamdagni, R. and Kebarle, P. (1973). *J. Amer. Chem. Soc.*, **95**, 3504
105. Dzidic, I. (1972). *J. Amer. Chem. Soc.*, **94**, 8333
106. Henderson, W. G., Taagepera, M., Holtz, D., McIver, R. J. Jr., Beauchamp, J. L. and Taft, R. W. (1972). *J. Amer. Chem. Soc.*, **94**, 4728
107. Arnett, E. M., Jones, F. M. III, Taagepera, M., Henderson, W. G., Beauchamp, J. L., Holtz, D. and Taft, R. W. (1972). *J. Amer. Chem. Soc.*, **94**, 4724
108. Aue, D. H., Webb, H. M. and Bowers, M. T. (1972). *J. Amer. Chem. Soc.*, **94**, 4726
109. Aue, D. H., Webb, H. M. and Bowers, M. T. (1973). *J. Amer. Chem. Soc.*, **95**, 2699
110. Long, J. and Munson, B. (1973). *J. Amer. Chem. Soc.*, **75**, 2427
111. Morton, T. H. and Beauchamp, J. L. (1972). *J. Amer. Chem. Soc.*, **94**, 3671
112. Beauchamp, J. L., Holtz, D., Woodgate, S. D. and Patt, S. L. (1972). *J. Amer. Chem. Soc.*, **94**, 2800
113. Brauman, J. I., Eyler, J. R., Blair, L. K., White, M. J., Comisarow, M. B. and Smyth, K. C. (1971). *J. Amer. Chem. Soc.*, **93**, 6360
114. McIver, R. T. Jr. and Silver, J. H. (1973). *J. Amer. Chem. Soc.*, **95**, 8462
115. Yamdagni, R. and Kebarle, P. (1973). *J. Amer. Chem. Soc.*, **95**, 4050
116. Brauman, J. I., Lieder, C. A. and White, M. J. (1973). *J. Amer. Chem. Soc.*, **95**, 927
117. McMahon, T. B., Blint, R. J., Ridge, D, P. and Beauchamp, J. L. (1972). *J. Amer. Chem. Soc.*, **94**, 8935; (1974). *J. Amer. Chem. Soc.*, **96**, 1269
118. Dawson, J. H. J., Henderson, W. G., O'Malley, R. M. and Jennings, K. R. (1973). *Int. J. Mass Spectrom. Ion Phys.*, **11**, 61
119. Bennett, S. L. and Field, F. H. (1972). *J. Amer. Chem. Soc.*, **94**, 5186
120. Cunningham, A. J., Payzant, J. D. and Kebarle, P. (1972). *J. Amer. Chem. Soc.*, **94**, 7627
121. Yamdagni, R. and Kebarle, P. (1971). *J. Amer. Chem. Soc.*, **93**, 7139
122. Grimsrud, E. and Kebarle, P. (1973). *J. Amer. Chem. Soc.*, **95**, 7939
123. Beauchamp, J. L. and Caserio, M. C. (1972). *J. Amer. Chem. Soc.*, **94**, 2638
124. Ridge, D. P. and Beauchamp, J. L. (1974). *J. Amer. Chem. Soc.*, **96**, 637
125. Tiedemann, P. W. and Riveros, J. M. (1974). *J. Amer. Chem. Soc.*, **96**, 185
126. Dunbar, R. C., Shen, J. and Olah, G. A. (1972). *J. Amer. Chem. Soc.*, **94**, 6862
127. Dunbar, R. C., Shen, J., Melby, E. and Olah, G. A. (1973). *J. Amer. Chem. Soc.*, **95**, 7200
128. Meot-Ner, M. and Field, F. H. (1973). *J. Amer. Chem. Soc.*, **95**, 7207
129. Winkler, H. U. and Beckey, H. D. (1972). *Org. Mass Spectrom.*, **6**, 655
130. Weygand, F. and Obermeier, R. (1971). *Eur. J. Biochem.*, **20**, 72
131. Sun, T. and Lovins, R. E. (1972). *Org. Mass Spectrom.*, **6**, 39
132. Sun, T. and Lovins, R. E. (1972). *Anal. Biochem.*, **45**, 176
133. Fairwell, T., Ellis, S. and Lovins, R. E. (1973). *Anal. Biochem.*, **53**, 115
134. Marino, G., Valente, L., Johnstone, R. A. W., Mohammedi-Tabrizi, F. and Sodini, G. C. (1972). *J. Chem. Soc. Chem. Commun.*, 357
135. Leclerq, P. A. and Desiderio, D. M. (1971). *Biochem. Biophys. Res. Commun.*, **45**, 308
136. Polan, M. L., McMurray, W. J., Lipsky, S. R. and Lande, S. (1972). *J. Amer. Chem. Soc.*, **94**, 2847
137. Leclerq, P. A., Desiderio, D. M. and Smith, L. C. (1971). *Biochem. Biophys. Res. Commun.*, **45**, 937

138. Patil, G. V., Hamilton, R. E. and Day, R. A. (1973). *Org. Mass Spectrom.*, **7**, 817
139. Morris, H. R., Williams, D. H. and Ambler, R. P. (1971). *Biochem. J.*, **125**, 189
140. Morris, H. R. and Williams, D. H. (1972). *J. Chem. Soc. Chem. Commun.*, 114
141. Wipf, H.-K., Irving, P., McCamish, M., Venkataraghavan, R. and McLafferty, F. W. (1973). *J. Amer. Chem. Soc.*, **95**, 3369
142. Winkler, H. U. and Beckey, H. D. (1972). *Biochem. Biophys. Res. Commun.*, **46**, 391
143. Nau, H., Kelley, J. A. and Biemann, J. (1973). *J. Amer. Chem. Soc.*, **95**, 7162
144. Polan, M. L., McMurray, W. J., Lipsky, S. R. and Lande, S. (1972). *J. Amer. Chem. Soc.*, **94**, 2847
145. Rostovtseva, L. I., Kiryushkin, A. A. and Khokhlov, A. S. (1971). *Zh. Obshch. Khim.*, **41**, 1380
146. Rostovtseva, L. I. and Kiryushkin, A. A. (1972). *Org. Mass Spectrom.*, **6**, 1
147. Bogentoft, C., Chang, J.-K., Sievertsson, H., Currie, B. and Folkers, K. (1972). *Org. Mass Spectrom.*, **6**, 735
148. Beuhler, R. J., Flanigan, E., Greene, L. J. and Friedman, L. (1972). *Biochem. Biophys. Res. Commun.*, **46**, 1082
149. Compernolle, F., Venderhaaghe, H. and Janssen, G. (1972). *Org. Mass Spectrom.*, **6**, 151
150. Kamerling, J. P., Vliegenthart, J. F. G., Vink, J. and de Ridder, J. J. (1971). *Tetrahedron Lett.*, 2367
151. Das, K. G. and Thayumanavan, B. (1972). *Org. Mass Spectrom.*, **6**, 1063
152. Vink, J., Bruins Slot, J. H. W., de Ridder, J. J., Kamerling, J. P. and Vliegenthart, J. F. G. (1972). *J. Amer. Chem. Soc.*, **94**, 2542
153. Schulten, H.-R., Beakey, H. D., Bessell, E. M., Foster, H. B., Jarman, M. and Westwood, J. H. (1973). *J. Chem. Soc. Chem. Commun.*, 416
154. Lehmann, W. D., Schulten, H.-R. and Beckey, H. D. (1973). *Org. Mass Spectrom.*, **7**, 1103
155. McCloskey, J. A., Futrell, J. H., Elwood, T. A., Schramm, K. H,, Banzica, R. P. and Townsend, L. B. (1973). *J. Amer. Chem. Soc.*, **95**, 5762
156. Westmore, J. B., Lin, D. C. K., Ogilvie, K. K., Wayborn, H. and Berestianski, J. (1972). *Org. Mass Spectrom.*, **6**, 1243
157. Schulten, H.-R. and Beckey, H. D. (1973). *Org. Mass Spectrom.*, **7**, 861
158. Mitscher, L. A., Showalter, H. D. H. and Foltz, R. L. (1972). *J. Chem. Soc. Chem. Commun.*, 796
159. McLeod, J. K. and Wells, R. J. (1972). *J. Amer. Chem. Soc.*, **95**, 2387
160. Green, M. M., McGrew, J. G. II, and Moldowan, J. M. (1971). *J. Amer. Chem. Soc.*, **93**, 6700
161. Karpati, A., Rave, A., Deutsch, J. and Mandelbaum, A. (1973). *J. Amer. Chem. Soc.*, **95**, 4244
162. Biros, F. J., Dougherty, R. C. and Dalton, J. (1972). *Org. Mass Spectrom.*, **6**, 1161
163. Hammerum, S. and Djerassi, C. (1972). *J. Amer. Chem. Soc.*, **95**, 5806
164. Pechine, J. M. (1972). *Org. Mass Spectrom.*, **6**, 805
165. Gonnelle, Y., Pechine, J. M. and Solgadi, D. (1973). *Org. Mass Spectrom.*, **7**, 1287
166. Kim, J. K., Findlay, M. C., Henderson, W. G. and Caserio, M. C. (1973). *J. Amer. Chem. Soc.*, **95**, 2184
167. Dias, J. R. and Djerassi, C. (1972). *Org. Mass Spectrom.*, **6**, 385
168. White, E., Tsuboyama, V. S. and McCloskey, J. A. (1971). *J. Amer. Chem. Soc.*, **93**, 6340

2
Ultraviolet and Visible Spectroscopy

C. J. TIMMONS
University of Nottingham

2.1   INTRODUCTION   35
      2.1.1   *General comments*   36
      2.1.2   *The literature*   36
      2.1.3   *Experimental methods*   37

2.2   DETERMINATION OF CONSTITUTION   39
      2.2.1   *Isolated multiple bond chromophores*   39
      2.2.2   *Conjugated systems*   40
      2.2.3   *Aromatic systems*   45
              2.2.3.1   *Benzene derivatives*   45
              2.2.3.2   *Quinones and tropolone*   46
              2.2.3.3   *Polycyclic compounds*   46
      2.2.4   *Heterocyclic systems*   47
              2.2.4.1   *N-containing rings*   47
              2.2.4.2   *O- and S-containing rings*   49
              2.2.4.3   *Macrocyclic pigments*   49

2.3   MISCELLANEOUS TOPICS   50

2.1   INTRODUCTION

This chapter describes the continuing contribution that ultraviolet and visible spectroscopy makes in structure determination. The work reported was published mainly during 1972 and 1973. This review should be considered as a sequel to the earlier one[1].

2.1.1   General comments

In spite of the increased use of other spectroscopic techniques, particularly [13]C n.m.r. spectroscopy, electronic absorption spectroscopy is still a much studied technique for a number of reasons. Since our fundamental under- standing of electronic absorption spectra is far from complete, the more experimental data available the better is the chance of interpreting the spec- trum of an unknown compound satisfactorily. An increasing number of theoretical calculations of spectra are now being carried out as a result of both developments in theory and in ease and availability of computer facilities. The results of such calculations are being compared very carefully with observed spectra. Such comparisons can increase our understanding of spectra, particularly by making the assignment of bands more accurate. Such assignments are becoming more important in interpreting photochemical reactions. From the cost point of view, a simple ultraviolet and visible spectrophotometer is still the cheapest type of spectrometer on the market, so that many small laboratories can afford one whilst an n.m.r. or mass spectrometer may be too expensive. The quantitative precision and accuracy of the technique readily lends itself to assay purposes.

Whilst searches continue to be made for correlations between substituent effects and absorption band positions and intensities, difficulties arise in doing this quantitatively both as regards interpretation and accuracy of measurement[2]. The band maximum does not always represent the 0–0 transition and further is not always related in the same way to the 0–0 transition. Observed spectra often consist of overlapping bands so that the observed maxima may differ even further from the maxima of individual component bands. Quantitatively it might be expected that shifts in band positions due to substituent effects could be described by linear free energy relationships of the type:

$$\tilde{v} - \tilde{v}_0 = \rho\sigma$$

It should be noted in this connection that the wavenumber difference of the bands should be used rather than the wavelength difference since only the former is directly related to the energy difference. A recent study[2] has shown that some of the earlier claimed correlations between band positions and substituent parameters (σ) are unfounded, particularly when there is strong interaction between the substituent and the chromophoric system.

2.1.2   The literature

The best key to spectroscopic data is the continuing series of *Organic Electronic Spectral Data*. From Volume 8 onwards each volume covers the spectra published in one year. As Volume 9 covers 1967 there is a time lag of six years[3]. The *C.R.C. Atlas of Spectral Data and Physical Constants for Organic Compounds* is not an 'Atlas' in the sense of containing full curves of absorption spectra[4]. It lists alphabetically by name about 8000 compounds and for many it gives the coordinates of the ultraviolet absorption maxima

(λ, ε) in a stated solvent. It also contains a u.v. spectral index in which the data are sorted into groups on the basis of the intensity of the strongest band for each compound and within each group in decreasing order of wavelength. The biennial reviews in *Analytical Chemistry* provide a very useful guide to the recent literature. Both the reviews on ultraviolet spectrometry[5] and on light absorption spectrometry[6] describe tersely the work of the previous two years and list numerous references. In view of the extensive nature of these reviews, this chapter concentrates instead on representative examples of work particularly relevant in a general way to structure determination. Some aspects of the documentation of spectroscopy and retrieval of the information have been discussed in the *UV Spectrometry Group Bulletin*[7], which is a continuation of the former *Photoelectric Spectrometry Group Bulletin* which contained a cumulative index in its last number[8].

An *Atlas of Protein Spectra in the Ultraviolet and Visible Regions* has been published[9]. The *Specialist Periodical Reports* series has issued the first two volumes entitled *Molecular Spectroscopy*[10], but these include little information relevant to the use of electronic spectra for structure determination. Some of the papers from the 1972 International Conference on Molecular Spectroscopy at Wrocław have been published[11]. Several new books deal at any rate in part with structure determination using u.v. spectra[12-17]. New editions of other books[18,19] and of a programmed text[20] have been issued. Photochromism is the subject of a book[21]. A paper describes some experimental exercises in which compounds are prepared and their structures determined using u.v. and other spectra[22].

2.1.3   Experimental methods

Some trends are discernible in new instruments on the market[5,6,8,23]. Improvements in optics and electronics have allowed higher precision and accuracy at both low and high absorbances. Careful choice of the sample position and of the photomultiplier has enabled better results to be obtained on samples with high light scattering properties, such as turbid solutions. Digital output of the absorbance is particularly useful when extensive manipulation of the data is intended.

Difference spectroscopy, where the difference between the absorbances of two samples either of slightly different composition or of different physical state is measured, can provide useful information. Applications of this technique include the demonstration of the ionisation of a chromophore and the study of conformations of proteins[24]. Spectrophotometer requirements for difference measurements are more stringent than for conventional absorbance measurements.

Whilst dual wavelength spectroscopy was first used by Chance about 25 years ago[25], only recently have commercial instruments with this capability been available[23,26]. The sample is irradiated with two beams of monochromatic light of wavelengths λ_1 and λ_2. Some function of the absorbances at the two wavelengths can then be measured. The single sample can be used as its own reference and this has advantages, particularly for turbid samples.

By making the difference in the wavelength ($\lambda_1 - \lambda_2$) constant, first derivative spectra can then be measured[27,28]. Such spectra can be useful in detecting shoulders on the sides of major bands which might otherwise be overlooked.

Errors in spectrophotometric measurements arising from multiple reflections in cells have been shown to be quite small under normal conditions[29]. The traditional assumption that optimum measurement precision of absorbance occurs at 0.43 has been shown to be grossly in error under certain conditions[30].

Further attention has been paid to the far or vacuum ultraviolet region, but it is not clear whether there is much worthwhile there for structure determination[31-35].

It was stated[1] that the molar absorption coefficient (ε) is usually measured in units of l mol^{-1} cm^{-1}. This is still so. However, in a recent book the use of the units m^2 mol^{-1} has been advocated and used throughout the book[36]. Thus it is now imperative that the units used should always be given explicitly since the two alternatives differ by a factor of 10. In certain circumstances it may not be possible to make measurements around the maximum of the band concerned but only in a region away from the maximum. A method has been reported for computing the position and intensity of the maximum from measurements at three positions away from the maximum and assuming a Gaussian curve[37]. This method has been applied to the determination of the maximum in the spectra of proteins arising from the peptide bond which occurs in the region around 197 nm (50 800 cm^{-1}) which is instrumentally rather inaccessible. In this way an estimation of the protein concentration and the number of peptide bonds may be possible.

In many measured spectra the individual bands are not always well separated and instead a number of overlapping bands occur, together with some shoulders or points of inflection rather than clear maxima. Correlations using the observed maxima under these circumstances can be error. It is preferable to resolve the component bands mathematically and extract the positions and intensities of the component maxima. A computer program has been developed for this analysis on the basis of simulating the observed spectrum using a sum of asymmetric Gaussian curves[38]. Another program has been reported which can be used to check such deconvolution techniques and to synthesise a spectrum from calculated transition moments and energies[39].

For the identification of electronic transitions, polarisation measurements in ordered media are necessary. These have often been made on molecules incorporated in stretched films. This method has been applied to benzoic and monosubstituted benzoic esters of some steroids in polyethylene films and has also given information on the conformations of the molecules[40]. The bands of Orange II (the sodium salt of benzene-(1-azo-4-sulpho)-2-oxynaphthalene) in stretched poly(vinyl alcohol) have been investigated[41]. A new method of orienting the molecules is to apply an electric field to a solution of the compound in a liquid crystal mixture[42]. A theory for the treatment of such measurements has been developed. Using this method with *trans*-1,6-diphenylhexa-1,3,5-triene, a previously predicted but unobserved hidden transition has been recognised.

2.2   DETERMINATION OF CONSTITUTION

2.2.1   Isolated multiple bond chromophores

The most studied isolated multiple bond chromophore is the carbonyl group[1]. The special case of cyclopropyl ketones shows several bands which have been assigned as shown in Table 2.1[43]. Compared with non-cyclopropyl ketones they show an extra band at 55 000 cm^{-1}. Band positions for some saturated lactones have been compared with those for some unsaturated lactones and related compounds[44].

Table 2.1   Absorption bands of cyclopropyl ketones

Approx. $\tilde{v}_{max}$/cm^{-1}	Approx. λ_{max}/nm	Assignment
33 000	300	N→Π*
50 000	200	N→σ*, (N→H*)
55 000	180	Δ$_2$→Π*, CT absorption peculiar to cyclopropyl ketones
59 000	170	σ→Π*

Measurements and calculations have been made on some simple nitrate esters[45]. The tail of the long wavelength absorption around 38 000 cm^{-1} (263 nm) is due to an n→π_N^* transition. Probably other lower intensity unresolved bands are also present. These may be of importance in interpreting photochemical reactions.

Table 2.2   n→π* bands of some azo compounds

Compound	$\tilde{v}_{max}$/cm^{-1}	λ_{max}/nm	ε_{max}/l mol^{-1} cm^{-1}	Structure
trans-MeN=NMe	28 000	357	13	none
cis-MeN=NMe	28 400	352	240	none
(1)	28 000	357	60	some
(2)	27 000	370	150	sharp
(3)	29 500	339	400	sharp
(4)	25 900	386	770	—
trans-PhN=NPh	22 300	448	407	none
cis-PhN=NPh	23 000	435	1470	none

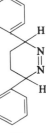

(1)   (2)   (3)   (4)

Spectroscopic properties of azo compounds have been reviewed with regard both to the available data and to the possible theoretical models[46]. They are characterised by a low lying n→π* transition which exhibits some unusual properties. It is influenced more by the local symmetry of the azo group than by the molecular geometry. For the *trans* isomers the transition is forbidden, but in some cases its unexpectedly high intensity, which may exceed that of the corresponding allowed transition in a closely related *cis* molecule, arises because it 'steals' intensity from the neighbouring π→π* transition. For most azo compounds the n→π* is devoid of vibrational fine structure. Some typical examples are listed in Table 2.2.

2.2.2 Conjugated systems

Although the observed spectra for the simpler conjugated systems have been well worked over, recent papers have dealt with the less common systems. New measurements have been made on cyclohexa-1,3-diene, *E*- and *Z*-hexa-1,3,5-triene. The assignments of the transitions are of importance for the interpretation of the photochemistry of these and related compounds[47].

The effect of a β-silyl group on the n→π* bands of some αβ-unsaturated ketones is to shift the band to a longer wavelength[48]. Some typical results are shown in Table 2.3, from which it can be seen that the shift in the dienone system is roughly twice the shift in the monenone system, which amounts to

Table 2.3   n→π* bands of some β-silyl substituted αβ-unsaturated ketones and of their carbon analogues in CCl$_4$

Compound	$\tilde{\nu}_{max}/cm^{-1}$	λ_{max}/nm	$\varepsilon_{max}/l\ mol^{-1}\ cm^{-1}$	$\Delta\tilde{\nu}_{max}/cm^{-1}$
(5)	26 800	372	16	2000
(6)	28 800	347	15	
(7)	26 700	374	19	1800
(8)	28 600	350	25	
(9)	28 600	350	35	800
(10)	29 400	340	26	
(11)	29 600	338	49	700
(12)	30 300	330	22	

(5)   (6)   (7)   (8)

(9)   (10)   (11)   (12)

700 cm^{-1} or *ca.* 2 kcal mol^{-1}. The effect of the β-silyl centre on the n→π* transition in αβ-unsaturated ketones is much smaller than the stabilisation by silicon in α-silyl ketones by a factor of *ca.* 10.

The measured spectra for a number of nitroethylenes have been compared with the results of calculations[49]. The spectra of a series of substituted oxamides, including 2,3-diketopiperazine, have been measured and the results correlated with the dihedral angle between the two carbonyl groups[50].

Although the general spectroscopic behaviour of dienes is well understood, that of nitrogen analogues such as the azines has been puzzling. Recent calculations[51] have clarified the picture. In dienes the only transition concerned is the π→π*. Relative rotation of the two double bonds separates the levels further with the consequent shift of the band to a higher wavenumber (shorter wavelength) and decreased intensity. With, for example, formaldehyde azine (13) there are additional levels to consider, an n level higher than

$$R^1R^2C=N-N=CR^1R^2$$

(13) $R^1 = R^2 = H$; (14) $R^1 = H$, $R^2 = Me$; (15) $R^1 = R^2 = Me$

the π level and a second π* level (π*b_g) higher than the π*a_u level which corresponds to the π* level in butadiene. These levels are shown to the left in Figure 2.1.

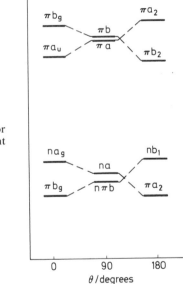

Figure 2.1 Energy levels for formaldehyde azine at different angles of twist

These levels would be expected to give rise to an intense π→π* (b_g→a_u) band and a lower-energy low-intensity n→π*a_u band. The b_g→b_g transition would be forbidden and the alternative n→π* (n→b_g) band would be masked by the intense π→π* band. This description assumes a planar *s-trans* ($\theta = 0°$) arrangement of the molecule. On twisting the molecule about the central bond, the two occupied levels move closer to each other and similarly the two unoccupied levels more towards each other. This causes the low-intensity

$n \rightarrow \pi^* a_u$ band to move to a higher wavenumber and the high-intensity band to be the envelope of contributions from the $n \rightarrow \pi^* b_g$ transition as well as from the $\pi b_g \rightarrow \pi^* a_u$ transition. Thus the high intensity band has the character of $n\pi \rightarrow \pi^*$ transitions. When the twist is through 90° all the transitions will have nearly similar energies. The overall effect of twisting should then be to affect the appearance of the spectrum and the number of overlapping bands, but the main high-intensity component should not be very sensitive to changes. Further, the extent of the twisting should be apparent from the spectrum.

Experimentally the azines (13)–(15) show several bands as listed in Table 2.4. All the bands show a shift to higher wavenumbers on changing from

Table 2.4   Light absorption properties* of some azines[51, 52]

Compound	$\tilde{v}_{max}/cm^{-1}$	λ_{max}/nm	$\varepsilon_{max}/l\ mol^{-1}\ cm^{-1}$	Comments
(13)†	34 500	289	35	observed bands
	44 150	227	275	
	51 700	193	8700	
(14)‡	35 200	284	89	observed bands
	44 400	225	1510	
(14)‡	34 200	292	76	components
	37 800	264	50	
	43 500	230	1510	
(15)‡	34 600	289	132	observed bands
	44 000	227	5010	
(15)‡	32 000	313	81	components
	35 200	284	123	
	43 200	231	3160	

* approximate values; † vapour phase; ‡ in cyclohexane

cyclohexane to ethanol, indicating at any rate some $n \rightarrow \pi^*$ character in the transitions involved. Resolution of the spectra of (14) and (15) into Gaussian components which may correspond to the individual transitions clarifies the interpretation. It is apparent that the spectra correspond to a non-planar conformation (Figure 2.2) rather than to the planar *s-trans* form. Calculations suggest that the non-planarity increases on going from (13) to (15).

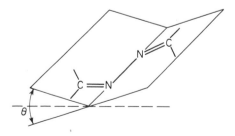

Figure 2.2   Non-planar conformation of azines

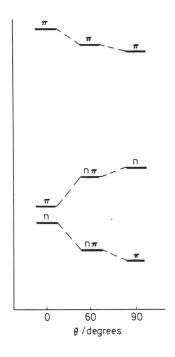

Figure 2.3 Energy levels for hydrazones at different angles of twist

Table 2.5   Light absorption properties of some hydrazones (16)[53]

$$
\begin{array}{c}
R^1 \quad R^2 \\
\diagdown \quad \diagup \\
\mathrm{N} \\
\diagdown_{\mathrm{N}} \diagup^{R} \\
\big| \\
\mathrm{Me}
\end{array}
$$

(16)

R^1	R^2	R^3	$\tilde{\nu}_{max}/\mathrm{cm}^{-1}$	$\lambda_{max}/\mathrm{nm}$	$\varepsilon_{max}/\mathrm{l\ mol}^{-1}\ \mathrm{cm}^{-1}$	*Solvent*
Me	H	H	43 500	230	5000	ethanol
			42 900	233	—	cyclohexane
Me	Me	H	43 900	228	4500	ethanol
			43 700	229	4800	cyclohexane
$(CH_2)_4$		H	42 700	234	3500	ethanol
			42 400	236	—	cyclohexane
$(CH_2)_5$		H	42 600	235	5500	ethanol
			42 200	237	—	cyclohexane
Me	H	Me	42 000	238	6400	ethanol
			41 600	240	8100	cyclohexane
Me	Me	Me	37 300	268	960	ethanol
			36 500	274	1020	cyclohexane
$(CH_2)_4$		Me	37 000	270	1100	ethanol
			36 100	277	1200	cyclohexane
$(CH_2)_5$		Me	36 500	274	900	ethanol
			35 600	281	—	cyclohexane

Hydrazones are not strictly conjugated compounds but can conveniently be discussed here since they have some similarities with azines. There is the possibility of rotation about the N—N bond as in the azines. Calculations[51] show that the energy levels are arranged as on the left-hand side of Figure 2.3 in the planar conformation. The lowest energy transition is of the $\pi\to\pi^*$ type, but on twisting it shifts to lower energy and becomes $n\pi\to\pi^*$ in character. Twisting through 90° changes it to an $n\to\pi^*$ transition. The observed spectra are explicable on this basis as illustrated in Table 2.5. When the groups R^2 and R^3 are both other than hydrogen, there is a great difference from the cases where one of these groups is hydrogen. In the former case the intensities are much lower, showing that the transitions are $n\pi\to\pi^*$ but largely $n\to\pi^*$ so that the molecules are non-planar. In the latter case the bands are more intense, at higher energies, and show smaller shifts. The direction of the solvent shifts is typical of $n\to\pi^*$ bands. Thus, these molecules are less non-planar and the transitions are $n\pi\to\pi^*$ but largely $\pi\to\pi^*$. Hence when R^2 and R^3 are other than hydrogen the molecule is considerably more non-planar.

There is still much interest in the light absorption properties and function of the visual pigments and related compounds. Some aspects of the spectral properties of retinal (17) are only incompletely understood so that its properties have been compared with related chromophores[54], as shown in Table 2.6.

Table 2.6 Light absorption properties for retinal and related compounds in EPA (ether, isopentane, alcohol) solution at 77 K

Compound	$\tilde{\nu}_{max}/cm^{-1}$	λ_{max}/nm	$\varepsilon_{max}/l\ mol^{-1}\ cm^{-1}$	Comment
(17)	25 800	388	55 000	broad band
(18)	30 200	331	52 000	broad band
(19)	24 900	402	154 000	lowest energy component of band with fine structure
(20)	27 900	358	162 000	lowest energy component of band (^1B_u) with fine structure
	24 600	407	171	^1A_g

Data for pure axerophtene (18) have not previously been available. It is of interest because it has the same system of carbon–carbon double bonds as retinal but lacks the carbonyl group. Like retinal it shows a broad band. In contrast, anhydrovitamin A (19) and 2,10-dimethylundecapentaene (20) both show bands with very marked vibrational fine structure. The latter also shows an additional band arising from the low-lying ^1A_g state.

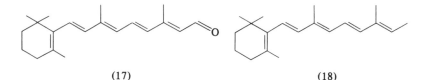

(17)   (18)

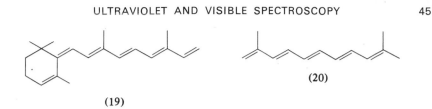

(19)   (20)

One of the important visual pigments is rhodopsin (21; RNH_2 is the protein opsin), closely related to retinal (17) but with the *cis* stereochemistry at the 11 double bond. One of the steps of the visual processes involves photobleaching of rhodopsin. It has now been confirmed by using difference spectra that this step is consistent with *cis–trans* isomerisation of the chromophore and that it cannot be attributed to changes in the conformation of the protein[55]. On the other hand, conformational analysis of the known crystal structures of compounds related to retinal suggests that various conformational states of the chromophore may contribute to the observed rhodopsin intermediates that have been characterised by different absorption maxima and may also contribute to the breadth of response in colour vision[56].

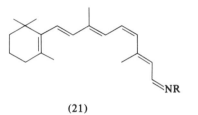

(21)

2.2.3   Aromatic systems

Much work has been published on various aspects of the electronic spectra of aromatic systems. In benzene itself the correct order of assignments of the transitions[57] is:

$$^3B_{1u} < {}^3E_{1u} < {}^1B_{2u} < {}^3B_{2u} < {}^1B_{1u} < {}^3E_{2g} < {}^1E_{1u} < {}^1E_{2g} < {}^3E_{2g}$$

The band at about 56 000 cm^{-1} is thus $^1E_{1u}$ and not $^1E_{2u}$ as has sometimes been stated. Clar has published a short book which summarises his theory and nomenclature for the spectra of aromatic molecules[58].

2.2.3.1   *Benzene derivatives*

Spectra are available for some azides, nitrenes and diazonium ions. These have been compared with results of calculations[59]. The spectra of *cis* and *trans* isomers of various β-substituted styrenes are interesting in that the conjugation band of the *cis* isomers is either at higher or lower energies than that of the *trans* isomers, depending on the nature of the β-substituent[60]. The isomer differences in energy were found to be linearly related to Taft's steric substituent constants. In all cases the intensity of the band of the *cis* isomer was smaller than that of the *trans*.

Comment has already been made about the breakdown of correlations between band positions and substituent constants for simple substituted benzenes, although some correlations for the intensities of the bands of some benzene derivatives are more promising[2]. A method has been proposed for the empirical determination, without resort to quantum mechanical calculation, of the number of bands and their origin in the spectra of tri-, tetra- and penta-substituted benzenes[61]. The method is applicable to derivatives in which the electron donor groups and the electron acceptor groups are situated in the *o*- and *p*-positions with respect to one another; if the donor group occupies the 1-position the other donor groups occupy odd positions while the acceptor groups occupy even positions of the ring. The method permits conclusions on the conformation of sterically hindered molecules on the basis of the absence or reduction in intensity of certain bands predicted for planar systems.

Several studies of nitro compounds have been made, including alkyl-substituted nitrobenzenes[62]. The empirical method[61] has been applied to nitroresorcinols[63]. The spectra of nitro derivatives of phenolic oligo nuclear compounds have been compared with related phenolic mononuclear compounds[64].

Other studies include methylphenols[65], hydroxy, methoxy and methylenedioxy derivatives of benzene, benzoic acid, benzaldehyde, acetophenone, cinnamic acid, etc.[66], dihydroxy- and diamino-benzenes[67], benzenethiols[68], arenethiols and alkyl aryl sulphides[69], methoxyphenyl thioethers[70], chlorinated biphenyls[71], solvent shifts of benzophenones[72], double-, triple- and quadruple-layered cyclophanes[73], aromatic azomethines[74,75], conjugated aromatic nitrones[76] and 1,2-diphenylcyclopropenyl cations[77].

2.2.3.2   *Quinones and tropolone*

A new book on quinones contains a section on their electronic spectra[78]. Studies have been reported on amino- and iodo-1,4-benzoquinones[79] and on polynuclear quinones[80]. Analysis of the near-u.v. spectrum of tropolone vapour has indicated that this band arises from a π-π^* transition rather than from an n$\rightarrow\pi^*$ transition and that the structure is typical of an aromatic molecule[81].

2.2.3.3   *Polycyclic compounds*

In the naphthalene series a comparison has been made between calculated and observed data for hydroxy and amino compounds[82]. Assignments for biphenylene have been discussed[83]. Dyes containing the phenalene ring system, e.g. (22), in conjugation with benzothiazole rings have been studied and the relative intensities shown to be closely related to the substitution pattern on the phenalene rings[84]. Some of the compounds absorb in the i.r.

region e.g. (22) has $\tilde{\nu}_{max}$ 11 000 cm^{-1}, λ_{max} 905 nm, ε_{max} 125 000 1 mol^{-1} cm^{-1}.

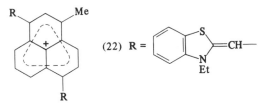

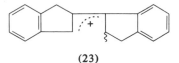

Earlier work on the triphenylmethane dyes has been extended to the naphthalene analogues of crystal violet, malachite green and Michler's hydrol blue[85]. The red colour observed in the cationic polymerisation of indene has been shown not to be due to the propagating species but is associated with the termination of polymerisation due to the ion (23)[86]. Several

(23)

series of linear conjugated systems (24) bearing aromatic terminal groups have been studied in detail[87].

$$R^1—(CH{=}CH)_n—R^2 \qquad R^1—(C{\equiv}C)_n—R^2$$

(24) R^1 and R^2 variously 1-anthryl, 2-pyrenyl, 2-fluorenyl, p-nitrophenyl, p-methoxyphenyl, 2-, 3- and 9-phenanthryl

2.2.4 Heterocyclic systems

Heterocyclic systems provide the most numerous examples of electronic spectra, but these spectra are some of the least well understood. Many of the correlations between spectra and structure are largely empirical. Recent advances in the techniques of quantum mechanical calculations, however, are now making interpretations of the spectra more firmly based on theory.

2.2.4.1 N-containing rings

New volumes in the series *Chemistry of Heterocyclic Compounds* contain sections on the spectra of pyridines[88], acridines[89] and pyridazines[90].

A new correlation has been described in which the cation of a 2-amino derivative of pyrimidine or of a fused pyrimidine has a spectrum similar to that of the neutral species of the 2-oxo analogue. This correlation, which does not extend to the 4-derivatives, is attributed to a formal similarity between the guanidinium and the urea types of resonance. The correlation was found to be less tight in the corresponding pyridine series, where amidinium and amide resonances were concerned[91].

Calculations have been made on the three isomeric pyridones and their
N-methyl derivatives. The results have been compared with the observed
spectra[92]. The unorthodox formula (25) has been proposed in the case of the

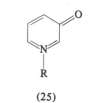

(25)

3-isomer. The spectra of the *N*-oxides of the pyridine-carboxylic acids and
-carboxamides have been measured and interpreted[93]. For the pyridinium
trihalides, diffuse reflectance spectra of the powders have been obtained[94].
Spectra have been discussed for amino derivatives of pyridine, pyrimidines,
acridines and quinolines[95], azanaphthalenes[96], 2-aminoquinoline and 4-
aminoquinaldine[97], benzo[*g*]quinoline derivatives[98] and some azo dyes containing the quinoxaline nucleus[99].

U.v. spectra can be used to distinguish 2- and 3-vinylindoles[100]. Measurement of the spectra in sulphuric acid (12.7 mol l^{-1}) at 30 and 70°C provides
a simple, quantitative and specific method for the identification of variously
substituted, naturally occurring plant indoles in μg quantities[101].

The spectra of some substituted 2,3-benzcarbazoloquinones have been
discussed[102]. 4-Nitro- and 4-nitroso-pyrazoles have been studied. The nitroso
compounds, unlike nitrosobenzenes, are mainly monomeric and thus are
blue or green due to the n→π* band[103]. The new ring system imidazo[1,2-*b*]-
pyrazole has been synthesised and the spectrum of the 6-methyl derivative
reported[104]. The spectra of some pyrazolo[5,1-*b*]quinazolones have been
measured and the n→π* bands indentified[105].

The structure of the antibiotic K16 (26) was elucidated partly by means of
its u.v. spectrum[106].

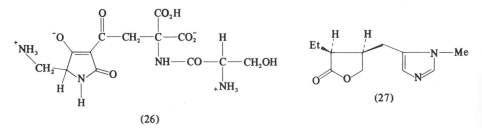

(26)   (27)

Studies on the imidazole alkaloid *d*-pilocarpine (27) show a characteristic
change in the spectrum on quaternisation. In water the alkaloid has a band
($\tilde{v}_{max}$ 46 500 cm^{-1}, λ_{max} 215 nm, ε_{max} 5780 l mol^{-1} cm^{-1}) due to the diene
system and a low intensity band ($\tilde{v}_{max}$ 40 000 cm^{-1}, λ_{max} 250 nm, ε_{max} 60
l mol^{-1} cm^{-1}) due to the n→π* transition of the C=N group. Quaternisation
causes the distinct higher intensity band to be replaced by a broad shoulder
(ε 3280 l mol^{-1} cm^{-1}) at about the same place. This shoulder has been

related to a shift due to the newly formed C—$\overset{+}{\text{N}}$ bond, causing a reduced delocalisation of the electron pair. No change in the low intensity band was observed[107].

Several studies of solvent effects have been made, e.g. on aminopyridines[108], diazines[109], *sym*-triazolo-4-*N*-acylimines[110] and triazolo-pyrazines and -pyrimidines[111].

Spectra are available for dibenz[*b*,*f*]azepines[112], benzimidazoles[113] and various methyl-substituted 11*H*-benzo[*a*]carbazoles[114].

2.2.4.2   *O- and S-containing rings*

Some data on O-containing ring systems are available in the new volume *Benzofurans*[115]. An *Atlas of Electronic Spectra of 5-Nitrofuran Compounds* has been published[116]. Assignments have been made for dibenzofuran[117]. In a study of some allylisoflavones it has been confirmed that, as in other cases, *O*-allylation causes a hypsochromic shift and *C*-allylation a slight bathochromic shift. The 8-allylisoflavones have only two maxima whereas the 6-allyl compounds have three. It was also checked that the isoflavanones and isoflavones show spectra closely resembling those of the acetophenones having the same substitution pattern[118]. Visible absorption spectra of some fluorescein derivatives have been measured in solutions of different pH and discussed in terms of various equilibria. The effect of substitution on the band position was noted[119].

For heterocyclic systems containing both O and N the results of calculations have been compared with observed spectra[120]. Spectra have been reported for furazan[121], isoxazolin-5-one[122], 2-isoxazolinium salts[123] and some *N*-aryl-2,3-dihydro-3-oxo-4*H*-1,4-oxazine derivatives[124].

The volume *Seven-Membered Heterocyclic Compounds Containing Oxygen or Sulphur* includes some uv spectra[125]. The spectra of thiophen and some deuteriated thiophens in the vapour phase have been measured and discussed[126]. Spectra are also available for some 2-thiophen carboxamides[127], dithieno- and furothieno-annelated tropones and tropylium ions[128], phenyl 2-thienyl ketone nitro derivatives[130], thiopyrylium and pyrylium cations[129] and pyrrolo[2,1-*b*]thiazole thioaldehydes[131].

2.2.4.3   *Macrocyclic pigments*

Electronic spectra are of prime importance in studies of the pyrrolic macrocyclic pigments such as porphyrins, chlorophylls and corrins. A complete volume of the *Annals of the New York Academy of Sciences* is entitled 'Chemical and Physical Behaviour of Porphyrin Compounds and Related Structures'[132]. It contains two reviews on porphyrin spectra[133,134]. Solution spectra of chlorophylls have been reviewed and some new assignments proposed[135]. A study of some porphyrins and their monocations and dications has confirmed that the neutral species and the corresponding mono- and di-protonated forms may be distinguished by their visible spectra. The acid ionisation leads either to monoanions with spectra similar to those of the corresponding monocations or to dianions with spectra similar to those of the corresponding dications. Spectra of two corroles and their protonated forms have also been reported[136,137]. In a series of papers on the influence of globin structure on the state of haeme, considerable use has been made of electronic spectra including difference spectra. The spectra of iron–porphyrin complexes and the shifts observed have been discussed[138].

2.3 MISCELLANEOUS TOPICS

In this section a few special topics including some less usual steric interactions are discussed.

The synthesis of the azo compound (28) and its unexpected spectrum (in cyclohexane, $\tilde{v}_{max}$ 37 300 and 22 400 cm^{-1}, λ_{max} 268 and 447 nm, ε_{max} 2100 and 150 l mol^{-1} cm^{-1}) together with the results of calculations have afforded a new correlation between the angle NNC subtended by the azo group and the energy of the n→π* transition. In (28) the unusually low energy is in keeping with the wide angle[139,140].

The special properties of cyclopropane rings are still being investigated[141-143]. Cyclopropenones have been reviewed[144]. The spectra of triptycenes including some heterocyclic triptycenes have been discussed[145]. Flavothebaone and its trimethyl ether have been known for a long time to exhibit abnormal spectra. Synthesis of some model compounds, e.g. (29), has confirmed the explanation that the medium intensity low-energy band arises from homoconjugation of the alkoxybenzene ring with the enone function. Solvent shifts show that it is a π→π* band[146].

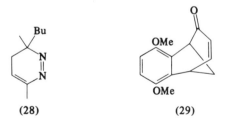

(28)	(29)

Spectra of dienes of the type (30) have been discussed with regard to the non-planarity of the systems[147]. The anomalous spectra of the bromination products (31) of 6,6-diphenylfulvene have been explained in terms of an interaction of the π-system with the bromine atoms at positions 1 and 4[148].

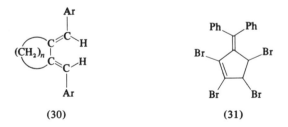

(30)	(31)

The light absorption properties of further various derivatives of carbonyl compounds have been described[149,150]. Methyl orange is used extensively as a probe for studying the interaction of organic anions with proteins, etc. The interaction is usually accompanied by spectral changes that are not associated with the acid–base indicator properties of the dye. It has been shown that the absorption curve consists of two overlapping bands which have been resolved mathematically. Consideration of the behaviour of the component bands shows that the observed shift in absorption maximum on binding of the dye to a protein should be interpreted as a shift in equilibrium rather than as a shift in transition energy[151].

Identification and determination of various water-soluble basic drugs by u.v. spectroscopy can be aided by conversion to their lauryl sulphates which can be extracted into organic solvents[152].

References

1. Timmons, C. J. (1973). *MTP International Review of Science, Organic Chemistry Series One*, Vol. 1, *Structure Determination in Organic Chemistry*, 63 (W. D. Ollis, editor) (London: Butterworths)
2. Brownlee, R. T. C. and Topsom, R. D. (1973). *Spectrochim. Acta*, **29A**, 385
3. Phillips, J. P., Feuer, H. and Thyagarajan, B. S. (1972). *Organic Electronic Spectral Data*, Vol. 8; (1973). *ibid.*, Vol. 9 (New York: Wiley-Interscience)
4. (1973). *C.R.C Atlas of Spectral Data and Physical Constants of Organic Compounds* (J. G. Grasselli, editor) (Cleveland: C.R.C. Press)
5. Hummel, R. and Kaufman, D. (1974). *Anal. Chem.*, **46**, 354R
6. Boltz, D. F. and Mellon, M. G. (1974). *Anal. Chem.*, **46**, 227R
7. (1973). *UV Spectrometry Group Bulletin*, No. 1 (Cambridge: UV Spectrometry Group)
8. (1972). *Photoelectric Spectrometry Group Bulletin*, No. 20 (Cambridge: Photoelectric Spectrometry Group)
9. (1972). *Atlas of Protein Spectra in the Ultraviolet and Visible Regions* (D. M. Kirschenbaum, editor) (New York: IFI/Plenum)
10. (1973–74). *Specialist Periodical Reports, Molecular Spectroscopy* Vols. 1, 2 (R. F. Barrow, D. A. Long and D. J. Millen, editors) (London: Chemical Society)
11. (1973). *J. Mol. Struct.*, **19**, 1
12. Parikh, V. M. (1974). *Absorption Spectroscopy of Organic Molecules* (Reading, Mass.: Addison-Wesley)
13. (1974). *An Introduction to Spectroscopic Methods for the Identification of Organic Compounds*, Vol. 2, (F. Scheinmann, editor) (Oxford: Pergamon)
14. Laszlo, P. and Stang, P. J. (1971). *Organic Spectroscopy: Principles and Applications* (New York: Harper and Row)
15. Simon, W. and Clerc, T. (1971). *Structural Analysis of Organic Compounds by Spectroscopic Methods* (London: Macdonald)
16. Brisdon, B. J. and Brown, D. W. (1973). *Structural Problems in Chemistry* (London: Van Nostrand Reinhold)
17. Dyer, J. R. (1972). *Organic Spectral Problems* (Englewood Cliffs, N.J.: Prentice-Hall)
18. Banwell, C. N. (1972). *Fundamentals of Molecular Spectroscopy*, 2nd ed. (London: McGraw-Hill)
19. Silverstein, R. M., Clayton, G. and Morrill, T. C. (1974). *Spectrometric Identification of Organic Compounds*, 3rd ed. (London: Wiley)
20. Cresswell, C. J., Runquist, O. A. and Campbell, M. M. (1972). *Spectral Analysis of Organic Compounds. An Introductory Programmed Text*, 2nd ed. (London: Longmans)
21. (1971). *Photochromism* (G. H. Brown, editor) (New York: Wiley-Interscience)
22. Jefford, C. W., McCreadie, R., Muller, P. and Pfyffer, J. (1973). *J. Chem. Educ.*, **50**, 177
23. (1974). *UV Spectrometry Group Bulletin*, No. 2 (Cambridge: UV Spectrometry Group)

24. Erickson, J. O. and Bramston-Cook, R. (1973). *Varian Instrument Applications*, **7**, (4), 10
25. Chance, B. (1951). *Rev. Sci. Instrum.*, **22**, 634
26. Porro, T. J. (1972). *Anal. Chem.*, **44**, 93A
27. Simonet, J.-C. (1972). *Analysis*, **1**, 183
28. Bramston-Cook, R. (1973). *Varian Instrument Applications*, **7**, (2), 11
29. Burnett, R. W. (1973). *Anal. Chem.*, **45**, 383
30. Ingle, J. D. and Crouch, S. R. (1972). *Anal. Chem.*, **44**, 1375
31. Bramston-Cook, R. and Erickson, J. O. (1973). *Varian Instrument Applications*, **7**, (3), 5
32. Fox, M. F. (1973). *Appl. Spectrosc.*, **27**, 155
33. Zaidel, A. N. and Shreider, E. Y. (1971). *Vacuum Ultraviolet Spectroscopy* (Jerusalem: Keter Publishing House)
34. (1974). *Chemical Spectroscopy and Photochemistry in the Vacuum-Ultra-Violet, Proceedings of the NATO Advanced Study Institute, held at Valmorin, Quebec*, **1973** (C. Sandorfy, P. Ausloos and M. B. Robin, editors) (Dordrecht: Reidel)
35. Fox, M. F. and Hayon, E. (1972). *J. Phys. Chem.*, **76**, 2703
36. Wells, C. H. J. (1972). *Introduction to Molecular Photochemistry* (London: Chapman and Hall)
37. Hocman, G. (1972). *Int. J. Biochem.*, **3**, 588
38. Klabuhn, B., Spindler, D. and Goetz, H. (1973). *Spectrochim. Acta*, **29A**, 1283
39. Price, E. R., Stoklosa, H. J. and Wasson, J. R. (1973). *J. Chem. Educ.*, **50**, 177
40. Yogev, A., Margulies, L. and Mazur, Y. (1971). *J. Amer. Chem. Soc.*, **93**, 249
41. Popov, K. R. (1972). *Opt. Spektrosk.*, **33**, 51
42. Cehelnik, E. D., Cundall, R. B., Timmons, C. J. and Bowley, R. M. (1973). *Proc. Roy. Soc., Ser. A*, **335**, 387
43. Meyer, A. Y., Muel, B. and Kasha, M. (1972). *J. Mol. Spectrosc.*, **43**, 262
44. Schepartz, A. I., Fleischman, R. A. and Cisle, J. H. (1972). *J. Chromatog.*, **69**, 411
45. Csizmadia, V. M., Houlden, S. A., Koves, G. J., Boggs, J. M. and Csizmadia, I. G. (1973). *J. Org. Chem.*, **38**, 2281
46. Rau, H. (1973). *Angew. Chem.*, **85**, 248
47. Minnaard, N. G. and Havinga, E. (1973). *Rec. Trav. Chim. Pays-Bas*, **92**, 1179
48. Felix, R. A. and Weber, W. P. (1972). *J. Org. Chem.*, **37**, 2323
49. Isaksson, G. and Sandström, J. (1973). *Acta Chem. Scand.*, **27**, 1183
50. Larson, D. B. and McGlynn, S. P. (1973). *J. Mol. Spectrosc.*, **47**, 469
51. Zverev, V. V., Él'man, M. S., Stolyarov, A. P., Ermolaeva, A. V. and Kitaev, Y. P. (1973). *Zh. Obshch. Khim.*, **43**, 2019
52. Ogilvie, J. F. and Horne, D. G. (1968). *J. Chem. Phys.*, **48**, 2248
53. Barltrop, J. A. and Conlong, M. (1967). *J. Chem. Soc. B*, 1081
54. Christensen, R. L. and Kohler, B. E. (1973). *Photochem. Photobiol.*, **18**, 293
55. Ebrey, T. G. and Honig, B. (1972). *Proc. Nat. Acad. Sci. USA*, **69**, 1897
56. Sundaralingam, M. and Beddell, C. (1972). *Proc. Nat. Acad. Sci. USA*, **69**, 1569
57. Karwowski, J. (1973). *J. Mol. Struct.*, **19**, 143
58. Clar, E. (1972). *The Aromatic Sextet* (New York: Wiley)
59. Kashiwagi, H., Iwata, S., Yamaoka, T. and Nagakura, S. (1973). *Bull. Chem. Soc. Jap.*, **46**, 417
60. Fueno, T., Yamaguchi, K. and Naka, Y. (1972). *Bull. Chem. Soc. Jap.*, **45**, 3294
61. Milliaresi, E. E., Ruchkin, V. E., Orlova, T. I. and Efremov, V. V. (1972). *Dokl. Akad. Nauk SSSR*, **205**, 353
62. Baas, J. M. A. and Wepster, B. M. (1972). *Rec. Trav. Chim. Pays-Bas*, **91**, 1002
63. Ruchkin, V. E. and Milliaresi, E. E. (1972). *Zh. Obshch. Khim.*, **42**, 2725
64. Böhmer, V., Deveaux, J. and Kämmerer, H. (1972). *Spectrochim. Acta*, **28A**, 1977
65. Baruah, G. D., Lal, B. B. and Singh, I. S. (1972). *Indian J. Pure Appl. Phys.*, **10**, 322
66. Šantavý, F., Walterová, D. and Hruban, L. (1972). *Coll. Czech. Chem. Commun.*, **37**, 1825
67. Yadav, J. S., Mishra, P. C. and Rai, D. K. (1972). *J. Mol. Struct.*, **13**, 253
68. Abu-Eittah, R. H. and Hillal, R. H. (1972). *Appl. Spectrosc.*, **26**, 270
69. Jones, I. W. and Tebby, J. C. (1973). *J. Chem. Soc. Perkin Trans. II*, 1125
70. Bermingham, G. E. and Smith, N. H. P. (1972). *Spectrochim. Acta*, **28A**, 1415
71. Dreeskamp, H., Hutzinger, O. and Zander, M. (1972). *Z. Naturforsch.*, **27a**, 756

72. Malawer, E. and Marzzacco, C. (1973). *J. Mol. Spectrosc.*, **46**, 341
73. Iwata, S., Fuke, K., Sasaki, M., Nagakura, S. and Otsubo, T. *J. Mol. Spectrosc.*, **46**, 1
74. Singh, N. and Krishan, K. (1973). *J. Indian Chem. Soc.*, **50**, 277
75. Dehne, H. and Rüttinger, H.-H. (1973). *Z. Chem.*, **13**, 235
76. Singh, N. and Krishan, K. (1973). *Indian J. Chem.*, **11**, 884
77. Yoshida, Z. and Miyahara, H. (1972). *Bull. Chem. Soc. Jap.*, **45**, 1919
78. Patai, S. (editor). (1974). *Chemistry of the Quinonoid Compounds* (New York: Wiley-Interscience)
79. Herre, W. and Wunderer, H. (1972). *Tetrahedron*, **28**, 5433
80. Titz, M. and Nepraš, M. (1972). *Coll. Czech. Chem. Commun.*, **37**, 2674
81. Alves, A. C. P. and Hollas, J. M. (1972). *Mol. Phys.*, **23**, 927
82. Suzuki, S. and Fujii, T. (1973). *J. Mol. Spectrosc.*, **47**, 243
83. Zanon, I. (1973). *J. Chem. Soc. Faraday Trans. II*, **69**, 1164
84. Elwood, J. K. (1973). *J. Org. Chem.*, **38**, 2430
85. Hallas, G., Paskins, K. N. and Waring, D. R. (1972). *J. Chem. Soc. Perkin Trans. II*, 2281
86. Prosser, H. J. and Young, R. N. (1972). *Eur. Polm. J.*, **8**, 879
87. Takeuchi, Y., Akiyama, S. and Nakagawa, M. (1972). *Bull. Chem. Soc. Jap.*, **45**, 3183; Takeuchi, Y., Yasuhara, A., Akiyama, S. and Nakagawa, M. (1973). *Bull. Chem. Soc. Jap.*, **46**, 909; (1973). *ibid.*, 2822; Takeuchi, Y., Akiyama, S. and Nakagawa, M. (1973). *ibid.*, 2828; Akiyama, S., Takeuchi, Y., Yasuhara, A. and Nakagawa, M. (1973). *ibid.*, 2830
88. (1974). *Pyridine and Its Derivatives, Supplement Part 2* (R. A. Abramovitch, editor) (New York: Wiley-Interscience)
89. (1973). *Acridines*, 2nd ed. (R. M. Acheson, editor) (New York: Wiley-Interscience)
90. (1973). *Pyridazines* (R. N. Castle, editor) (New York: Wiley Interscience)
91. Albert, A. and Taguchi, H. (1973). *J. Chem. Soc. Perkin Trans. II*, 1101
92. Cignitti, M. and Paoloni, L. (1972). *Theor. Chim. Acta*, **25**, 277
93. Caswell, L. R., Lee, F. C. and Creagh, L. T. (1972). *J. Heterocycl. Chem.*, **9**, 551
94. Gabes, W., Stufkens, D. J. and Gerding, H. (1973). *J. Mol. Struct.*, **17**, 329
95. Girault, G., Coustal, S. and Rumpf, P. (1972). *Bull. Soc. Chim. Fr.*, 2787
96. Baba, H. and Yamazaki, I. (1972). *J. Mol. Spectrosc.*, **44**, 118
97. Kovi, P. J., Capomacchia, A. C. and Schulman, S. G. (1972). *Anal. Chem.*, **44**, 1611
98. Kosheleva, L. I., Kozyreva, N. P. and Bekhli, A. F. (1972). *Khim. Geterotsikl. Soedin.*, 662
99. Sherif, S., Issa, R. M. and Aggour, S. S. (1972). *J. Prakt. Chem.*, **314**, 15
100. Bailey, A. S. and Seager, J. F. (1974). *J. Chem. Soc. Perkin Trans. I*, 763
101. Mollan, R. C., Harmey, M. A. and Donnelly, D. M. X. (1973). *Phytochemistry*, **12**, 447
102. Issa, R. M., Hamman, A. S. and Etaiw, S. H. (1972). *Z. Phys. Chem.* (*Leipzig*), **251**, 177
103. Habraken, C. L., Beenakker, C. I. M. and Brussee, J. (1972). *J. Heterocycl. Chem.*, **9**, 939
104. Elguero, J., Jacquier, R. and Mignonac-Mondon, S. (1973). *J. Heterocycl. Chem.*, **10**, 411
105. Bhujle, V. V., Seshadri, S. and Padhye, M. R. (1972). *Indian J. Chem.*, **10**, 613
106. Batelaan, J. G., Barnick, J. W. F. K., van der Baan, J. L. and Bickelhaupt, F. (1972). *Tetrahedron Lett.*, 3107
107. Ben-Bassat, A. and Lavie, D. (1972). *Isr. J. Chem.*, **10**, 385
108. Pelloni-Tamaş, V., Rădulescu, N. and Simionovici, R. (1972). *Rev. Roum. Chim.*, **17**, 931
109. Launay, G. and Wojtkowiak, B. (1973). *C.R. Acad. Sci., Ser. C*, **276**, 225
110. Timpe, H.-J. and Becker, H. G. O. (1972). *Chimia*, **26**, 473
111. Moxon, G. H. and Slifkin, M. A. (1972). *Spectrochim. Acta*, **28A**, 2419
112. Kricka, L. J. and Ledwith, A. (1974). *Chem. Rev.*, **74**, 101
113. Preston, P. N. (1974). *Chem. Rev.*, **74**, 279
114. Carruthers, W. and Evans, N. (1974). *J. Chem. Soc. Perkin Trans. I*, 421
115. (1974). *Benzofurans* (A. Mustafa, editor) (New York: Wiley-Interscience)
116. Eidus, J., Ekmane, A., Venters, K. and Hiller, S. (1970). *Atlas of Electron(ic) Spectra of 5-Nitrofuran Compounds* (Jerusalem: Keter Publishing House)
117. Bree, A., Vilkos, V. V. B. and Zwarich, R. (1973). *J. Mol. Spectrosc.*, **48**, 135

118. Krishnamoorthy, V., Seshadri, T. R. and Krishnaswamy, N. R. (1972). *Indian J. Chem.*, **10**, 258
119. Issa, I. M., Issa, R. M. and Ghoneim, M. M. (1972). *Z. Phys. Chem. (Leipzig)*, **250**, 161
120. Ferré, Y., Faure, R. and Vincent, E.-J. (1972). *J. Chim. Phys. Physiochim. Biol.*, **69**, 860
121. Forrest, B. J. and Richardson, A. W. (1972). *Can. J. Chem.*, **50**, 2088
122. Teysseyre, J., Sauvaitre, H., Arriau, J. and Deschamps, J. (1972). *J. Mol. Struct.*, **12**, 191
123. Belly, A., Jacquier, R., Petrus, C. and Petrus, F. (1973). *Bull. Soc. Chim Fr.*, 1390
124. Abdulla, R. F. (1973). *J. Heterocycl. Chem.*, **10**, 347
125. (1972). *Seven-Membered Heterocyclic Compounds Containing Oxygen or Sulphur* (A. Rosowsky, editor) (New York: Wiley-Interscience)
126. Di Lonardo, G., Galloni, G., Trombetti, A. and Zauli, C. (1972). *J. Chem. Soc. Faraday Trans. II*, **68**, 2009
127. Alberghina, G., Arcoria, A., Fisichella, S. and Scarlatta, G. (1972). *Spectrochim. Acta*, **28A**, 2063
128. Liljefors, T., Michael, U., Yom-Tov, B. and Gronowitz, S. (1973). *Acta Chem. Scand.*, **27**, 2485
129. Yoshida, Z., Sugomoto, H. and Yoneda, S. (1972). *Tetrahedron*, **28**, 5873
130. Arcoria, A., Maccarone, E., Musumarra, G. and Romano, G. (1973). *Spectrochim. Acta*, **29A**, 161
131. Mackie, R. K., McKenzie, S., Reid, D. H. and Webster, R. G. (1973). *J. Chem. Soc. Perkin Trans. I*, 657
132. Adler, A. D. (editor) (1973). *Ann. N.Y. Acad. Sci.*, **206**, 1
133. Corwin, A. H. (1973). *Ann. N.Y. Acad. Sci.*, **206**, 201
134. Treibs, A. (1973). *Ann. N.Y. Acad. Sci.*, **206**, 97
135. Weiss, C. (1972). *J. Mol. Spectrosc.*, **44**, 37
136. Grigg, R., Hamilton, R. J., Jozefowicz, M. L., Rochester, C. H., Terrell, R. J. and Wickwar, H. (1973). *J. Chem. Soc. Perkin Trans. II*, 407
137. Clarke, J. A., Dawson, P. J., Grigg, R. and Rochester, C. H. (1973). *J. Chem. Soc. Perkin Trans. II*, 414
138. Perutz, M. F., Ladner, J. E., Simon, S. R., Ho, C. (1974). *Biochemistry*, **13**, 2163; Perutz, M. F., Fersht, A. R., Simon, S. R. and Roberts, G. C. K. (1974). *ibid.*, **13**, 2174; Perutz, M. F., Heidner, E. J., Ladner, J. E., Beetlestone, J. G., Ho, C. and Slade, E. F. (1974). *ibid.*, **13**, 2187
139. Baird, N. G., de Mayo, P., Swenson, J. R. and Usselman, M. C. (1973). *J. Chem. Soc. Chem. Commun.*, 314
140. de Mayo, P. and Usselman, M. C. (1973). *Can. J. Chem.*, **51**, 1724
141. Van-Catledge, F. A. (1973). *J. Amer. Chem. Soc.*, **95**, 1173
142. Brown, R. S. and Traylor, T. G. (1973). *J. Amer. Chem. Soc.*, **95**, 8025
143. Martinelli, L., Mutha, S. C., Kethcam, R., Strait, L. A. and Cavestri, R. (1972). *J. Org. Chem.*, **37**, 2278
144. Potts, K. T. and Baum, J. S. (1974). *Chem. Rev.*, **74**, 189
145. de Wit, J. and Wynberg, H. (1973). *Tetrahedron*, **29**, 1379
146. Heaney, H., Hollinshead, J. H., Kirby, G. W., Ley, S. V., Sharma, R. P. and Bentley, K. W. (1973). *J. Chem. Soc. Perkin Trans. I*, 1840
147. Dale, J. and Kristiansen, P. O. (1972). *Acta Chem. Scand.*, **26**, 961
148. Cheng, P.-T., Gwinner, P. A., Nyburg, S. C., Stanforth, R. R. and Yates, P. (1973). *Tetrahedron*, **29**, 2699
149. Grammaticakis, P. and Boyer, R. (1972). *C. R. Acad. Sci., Ser. C*, **274**, 1941
150. Grammaticakis, P. (1972). *C. R. Acad. Sci., Ser. C*, **275**, 1431
151. Reeves, R. L., Kaiser, R. S., Maggio, M. S., Sylvestre, E. A. and Lawton, W. H. (1973). *Can. J. Chem.*, **51**, 628
152. Pellerin, F., Mancheron, D. and Demay, D. (1972). *Ann. Pharm. Fr.*, **30**, 429

3
Nuclear Magnetic Resonance Spectroscopy

I. O. SUTHERLAND
University of Sheffield

3.1 INTRODUCTION 56

3.2 ^{13}C N.M.R. 57
   3.2.1 *General comments* 57
   3.2.2 *Deuterium isotope effects* 58
   3.2.3 *Applications* 59
      3.2.3.1 *Assignments and general structural studies* 59
      3.2.3.2 *Cations and anions* 61
      3.2.3.3 *Biosynthesis and reaction mechanisms* 62

3.3 SHIFT REAGENTS 67
   3.3.1 *Lanthanide reagents* 67
      3.3.1.1 *Theory* 67
      3.3.1.2 *Applications* 71
   3.3.2 *Shift reagents other than lanthanides* 74
   3.3.3 *Chiral shift reagents, chiral solvents and related topics* 75

3.4 COUPLING CONSTANTS 78

3.5 STRUCTURAL STUDIES USING T_1 AND RELATED EFFECTS 81
   3.5.1 *T_1 measurements* 81
   3.5.2 *Nuclear Overhauser effect* 84

3.6 THE STUDY OF CONFORMATION AND CONFORMATIONAL CHANGES 86

3.1 INTRODUCTION

The period under review has been marked by the increasing use of Fourier transform (FT) n.m.r. spectrometers[1], particularly for ^{13}C n.m.r., and the widespread use of lanthanide shift reagents[2-5]. These two aspects of n.m.r. will receive considerable attention in this review. ^{1}H FT n.m.r. promises to be a particularly valuable technique for structure determination because of the enhanced sensitivity associated with the method, and it is encouraging to find that useful techniques such as homonuclear decoupling[6] and INDOR[7] can now be carried out in the FT mode as well as the more conventional continuous wave (CW) mode. The potential additional advantages associated with the computational aspects of FT n.m.r. have yet to be fully realised.

The application of the nuclear Overhauser effect (NOE) for structure determination, which dates from 1965, has received increased attention, particularly for conformational studies. The closely related measurement of spin–lattice relaxation times, T_1, is now made more readily using FT n.m.r. combined with suitable pulsing techniques[1], and T_1 measurements are beginning to be used for structure determination, particularly in ^{13}C n.m.r. Both of these techniques will be discussed together with the further application of T_1 measurements in studies with lanthanide reagents.

^{1}H N.m.r. spectra continue to provide most of the n.m.r. information that is used in structure determination in organic chemistry. Most aspects of this are well known and have been extensively documented; they will not therefore be covered in this review. The major advance in ^{1}H n.m.r. during the period 1972–73 has been the use of high operating frequencies (220–300 MHz) associated with the use of superconducting solenoids. To some extent the advantages of high frequencies are obtainable using lanthanide shift reagents, but the advantages of high-frequency spectrometers are particularly apparent for compounds that do not form lanthanide complexes and for conformational studies.

Studies of the conformations and conformational changes of organic compounds have also received considerable attention, techniques for obtaining spectra at very low temperatures have improved, and the complementary use of ^{1}H and ^{13}C n.m.r. has been very successful. This aspect of structural examination will be fully reviewed.

^{13}C and ^{1}H data dominate structural work in organic chemistry, but recent reviews on nitrogen n.m.r.[8] and phosphorus n.m.r.[9] indicate fully their more specialised use. Other specialised topics that have been reviewed during 1972–73 include the n.m.r. spectra of carbohydrates[10,11] the conformational analysis of heterocyclic compounds[12], the stereochemistry of double bonds[13], spectral analysis[14], and the use of computers in n.m.r. spectroscopy[15].

This review will be organised on the basis of the topics outlined above. All three aspects of structure determination—constitution, configuration and conformation—will be covered by the examples chosen to illustrate the use of these techniques. The choice of topics is necessarily restricted, owing to the large number of applications of n.m.r. spectroscopy, and it is not possible to cover two other important aspects of n.m.r. spectroscopy. These are the study of reaction mechanisms using CIDNP observations, and the use of

[1]H n.m.r. to detect ring current effects in cyclic unsaturated systems. Both topics have been developed and extended during the period under review.

3.2 [13]C N.M.R.

3.2.1 General comments

The proportion of papers using [13]C n.m.r. has increased markedly during the period under review but [1]H n.m.r. still remains the predominant technique for structure determination. The use of ^{13}C spectra[16-18] tends to be complementary to the use of ^{1}H spectra and the two techniques make a powerful combination for attacking many structural problems (see Section 3.2.3.1), particularly problems involving conformation (see Section 3.6) and the examination of charged systems (Section 3.2.3.2). The increasing importance of [13]C n.m.r. in biosynthetic studies, and potentially in other mechanistic studies involving C—C bond formation, is evident from the large number of papers that have appeared. The application of [13]C T_1 studies in structure determination is also currently under investigation (see Section 3.5.1.).

A number of techniques have been suggested for solving some of the problems inherent in FT [13]C n.m.r. The long relaxation times of quaternary carbons tend to lead to low-intensity signals in spectra that are accumulated using normal delay times (1–2 s). These long relaxation times may conveniently be shortened[19] by adding a paramagnetic relaxation reagent such as $Cr(acac)_3$ or $Fe(acac)_3$. This procedure leads to easier observation of the quaternary carbon resonances and an improvement in the quantitative intensity relationships of the ^{13}C spectrum. The comparison of intensities in the presence or absence of the relaxation reagent is expected[20] to give maximum structural information. The use of very low power for the broad-band ^{1}H decoupler, rather than the usual very high power required for fully decoupled ^{13}C spectra, provides an alternative technique[21] for identifying quaternary carbons. This leads to selective decoupling of the weakly coupled quaternary carbons which appear as sharp singlets above the low-intensity background.

Broad resonances in ^{13}C spectra may be filtered out by introducing a delay time between the end of the pulse and the start of data collection[22]. This technique is useful for the observation of sharp lines from small substrates, e.g. in the presence of broad lines from substances such as proteins.

[13]C Assignments have been made for some years by a variety of ^{1}H decoupling techniques, and a graphical method for the most efficient use of off-resonance decoupling has been described[23]. The positions of peaks in the ^{13}C spectrum are plotted as a function of the position of irradiation in the ^{1}H spectrum; the collapse of ^{13}C multiplets may then most readily be related to the irradiation of specific ^{1}H resonances. The method was illustrated by the assignment of the ^{13}C spectrum of NAD[+] (1).

The number of protons attached to a carbon atom has usually been determined by off-resonance decoupling, and, in general, [13]CH_3 is expected to give a quartet, [13]CH_2 a triplet and [13]CH a doublet. Under certain circumstances this does not apply, owing to virtual coupling.[24] Thus in the absence of

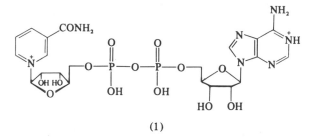

(1)

^{1}H decoupling the carbons of 1,2-dichloroethane give a triplet of triplets, which under off-resonance conditions becomes a complex multiplet rather than a simple triplet. Similarly, in the absence of ^{1}H decoupling the carbons of fumaric acid give a doublet of doublets which is changed to a five-line pattern in the off-resonance decoupled spectrum. Qualitatively these results are due to the comparable values of J_{CH} and J_{HH} under off-resonance conditions, and spectra may be reproduced by computation on that basis. A warning has been given[25] that the simple interpretation of pulsed spin decoupling experiments may also be misleading.

3.2.2 Deuterium isotope effects

A number of studies have appeared of deuterium isotope effects in ^{13}C n.m.r., and J_{CD} and ^{13}C chemical shifts have been listed[26] for a number of common deuteriated n.m.r. solvents. It has been known for some time that a-deuterium substitution, causes readily recognisable $^1J_{CD}$ coupling and specific shifts of the carbon resonance, e.g. upfield shifts for sp^3 hybridised carbon and downfield shifts for carbonyl carbon. In a study of long-chain compounds, upfield shifts were also detectable[27] for the β- and γ-carbons, together with increased linewidths due to long range C–D coupling. J_{CH}/J_{CD} is expected to be close to γ_H/γ_D (6.5144) and values lying in the range 6.49–6.61 have been found experimentally[28].

The use of deuterium labelling as a probe for ^{13}C assignments is not new, but the application of ^{13}C n.m.r. to detect deuterium labels and to measure deuterium distribution is a novel and promising aspect of ^{13}C n.m.r. as an aid to mechanistic studies[29,30]. Thus the analysis of deuterium scrambling between positions 1, 2 and 6 during the solvolysis of the deuteriated *exo*- and *endo*-norbornyl brosylates (2) may be based upon the ^{13}C spectrum of the product (3). The extent of deuterium labelling at each carbon atom of (3) is based upon integration of the isotopically shifted and coupled ^{13}C–D signal and the ^{13}C–H signal. In a more complex example, humulene (4) was treated with D_2SO_4 to give appollanol (5) as a mixture of the $^2H_0$–$^2H_5$ species; the total deuterium distribution over all fifteen carbon atoms of (5) was determined from the ^{13}C n.m.r. spectrum.

The homo-enolisation of fenchone (6), and resulting deuterium incorporation at C-6 and C-8, could also conveniently be followed[30] using ^{13}C n.m.r. spectroscopy, as were the relative rates of base-catalysed deuteriation at the aromatic carbon atoms of acenaphthene (7). The stereochemistry of the singlet

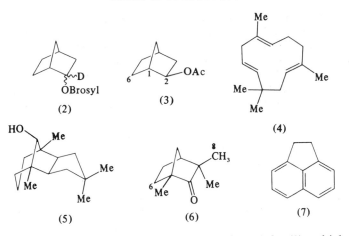

(2)   (3)   (4)

(5)   (6)   (7)

oxygen oxidation of the specifically deuteriated *R*-olefin (8), which gives a mixture of the *R*-alcohol (9) with no deuterium and the monodeuteriated *S*-alcohol (10), was established[31] by using the ^{13}C n.m.r. spectra of the diastereomeric (−)-methoxytrifluoromethylphenylacetates of (9) and (10). The

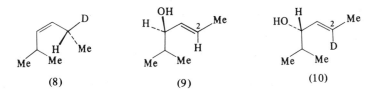

(8)   (9)   (10)

mixture of esters gives different C-2 resonances from the two diastereomers, and the appearance of these signals indicate the presence or absence of deuterium.

3.2.3   Applications

3.2.3.1   *Assignment and general structural studies*

Studies involving the assignment of the ^{13}C n.m.r. spectra of complex molecules of known structure, using the various standard methods for making assignments, continue to be carried out. The information obtained is relevant to structure determination of analogous compounds by ^{13}C n.m.r., and rules for predicting and assigning chemical shifts continue to be elaborated and improved. Two books[16,17], a compilation of spectral data[18] and a chart showing chemical shift and structure correlation[32] are now available to assist in the interpretation of ^{13}C spectra. Off-resonance and other decoupling techniques are widely used and well established; to these should be added the use of lanthanide shift reagents, the measurement of T_1 values, and the use of relaxation reagents such as $Cr(acac)_3$. A good illustration of the use of all of these techniques is provided by the investigation of cholic acid and deoxycholate–hydrocarbon complexes[33]. The possibility of computer-assisted identification of unknown compounds by ^{13}C n.m.r., based upon the additivity of substituent effects, has also been advocated[34]. The application of partially relaxed FT n.m.r. in structural studies will be discussed later (Section 3.5.1).

Surprisingly few studies have appeared in which ^{13}C n.m.r. is essential for the determination of structure. The diterpenoid alkaloid (11) was formulated largely on the basis of its ^{13}C n.m.r. spectrum, ^{13}C assignments having previously been made for closely related diterpene alkaloids of known structure[35].

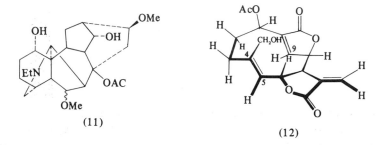

(11)

(12)

A more typical example is provided by the germacranolide sesquiterpene dilactone (12)[36], since this involved the use of both ^{13}C and ^{1}H n.m.r. data. The structure determination included the recognition of seventeen carbon atoms from the ^{13}C n.m.r. spectrum, assignable as $3 \times \underline{C}O{-}O$, $3 \times \underline{C}{=}$, $3 \times C{-}\underline{C}H{-}O$, $2 \times \underline{C}H{=}$, $2 \times C{-}\underline{C}H_2{-}C$, $\underline{C}H_2{=}$, $C{-}\underline{C}H_2{-}O$, $C{-}\underline{C}H{-}C$ and $-\underline{C}H_3$. The 300 MHz ^{1}H n.m.r. spectrum of (12) was consistent with these assignments and showed a number of coupling patterns that were analysed by double resonance experiments; in particular, the presence of the groupings (13) and (14) was established. This information, together

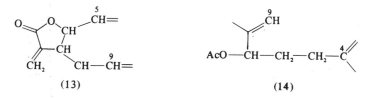

(13)

(14)

with biogenetic considerations and the observed values of vicinal coupling constants, led to the constitution, configuration and conformation shown in (12).

^{13}C N.m.r. shifts may also be used for establishing relative configuration in structural studies. The assignment of configuration at C-1 of the alcohol (15) was based[37] upon ^{13}C n.m.r., and the configuration at C-13 in the degradation

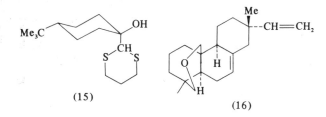

(15)

(16)

product (16) of annonolide was based[38], in part, upon the similarity of the C-13 shift with that of the analogous carbon in isopimaradiene. It has also been pointed out[39] that the ^{13}C shift of the 19-methyl group of steroids distinguishes between *cis*- and *trans*-A/B ring fusion.

3.2.3.2 Cations and anions

Study of the ^{13}C n.m.r. spectra of charged species, particularly cations, is rapidly becoming the most important established technique for structure determination. The original ^{13}C assignments for the norbornyl cation, which strongly support the 'non-classical' structure, were based upon INDOR derived spectra, but a more recent examination[40] using FT ^{13}C n.m.r. is in accord with the earlier assignments. Other highly 'non-classical' structures for carbonium ions have also been proposed on the basis of ^{13}C spectra, but in some cases it is difficult to exclude alternatives based upon equilibrating sets of less highly delocalised cations.

The $(CH)_5^+$ system has been examined theoretically and it may now be exemplified[41] by the cation derived from the alcohol (17). The ^{13}C spectrum of the cation is in accord with the symmetry of (18), and the bridging $C-CH_3$

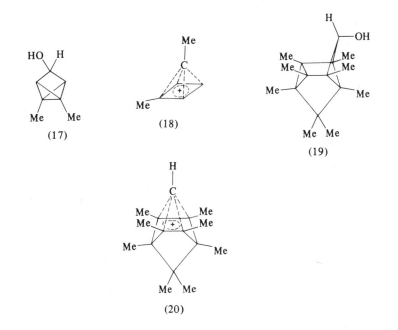

group shows an unusually high-field ^{13}C shift. The alcohol (19) gives a cation which may be the bis-homo derivative (20) of the same $(CH)_5^+$ system[42]; in this case ^{13}C n.m.r. again suggests that the positive charge is located at the base of the pyramid.

The homologous $(CH)_6^{2+}$ system is obtained[43] as its hexamethyl derivative

(21) from the alcohol (22); the dication (21) could be a mixture of equilibrating dications but the ^{13}C shifts of (21) are believed to be in accord with the
charge distribution expected for the indicated structure. The treatment of the
alcohol (23) with fluorosulphonic acid gives a cation ($C_{11}H_{11}^{+}$) which was
examined by ^{13}C and ^{1}H n.m.r.[44]. The spectra indicate that the cation has
five-fold degeneracy either as a result of a symmetrical structure or as the

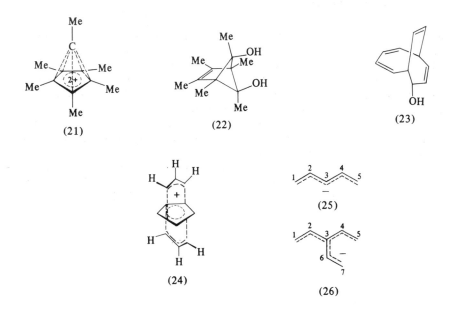

(21)   (22)   (23)

(24)   (25)

                        (26)

result of a suitable fluxional structure; arguments based upon the ^{13}C shifts
favour the highly delocalised, symmetrical '[3,5,3]armilenium' structure (24)
for the cation.

The charge distribution in carbanions may also be studied by ^{13}C n.m.r.[45],
and in delocalised anions such as (25) and (26) the odd-number carbon atoms
give higher field shifts (65–99 p.p.m.) than the even-number carbons (124–147
p.p.m.), in accord with the charge distribution based on HMO and other MO
treatments.

3.2.3.3   Studies of biosynthesis and reaction mechanisms

Studies of biosynthesis involving ^{13}C-enriched precursors followed by the
location of ^{13}C in metabolites by n.m.r. methods have been too numerous in
1972–73 to review in detail. A few selected examples will be discussed to
illustrate advances in scope and techniques.

The first example of ^{13}C utilisation for the examination of terpene biosynthesis has been reported, the diterpenes (27) being obtained from the
mushroom *Oospora virescens*. The ^{13}C spectra of (27) were assigned on the
basis of chemical shift theory and by comparison with reported assignments
for analogous diterpenes. The incorporation of [1-^{13}C]- and [2-^{13}C]-acetate

give the ^{13}C labelling patterns indicated in (27), which are in accord with current theories of terpene biosynthesis[46]. It is also of considerable interest that valine, with a stereospecific label in the isopropyl group, is converted in

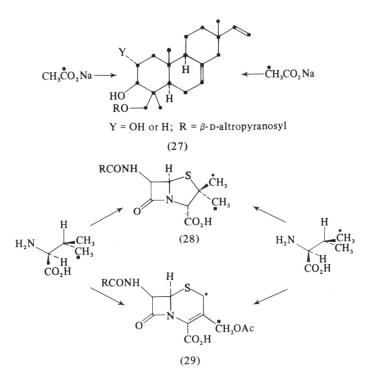

Y = OH or H; R = β-D-altropyranosyl

(27)

(28)

(29)

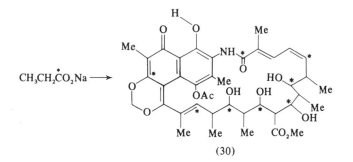

stereospecific fashion into penicillins (28) and cephalosporins (29)[47,48]. This result is particularly satisfying since it is based on the complementary work of two groups, each using a different ^{13}C label in the valine. The biosynthesis of streptovarian (30) was successfully studied[49] using [1-^{13}C]propionate,

(30)

which gives the labelling pattern shown. All of these studies indicate the efficiency of ^{13}C n.m.r. for biosynthetic studies, but probably the most impressive examples to date are provided by the related studies of porphyrin and Vitamin B_{12} biosynthesis. It is clearly not possible to do justice to this work in a brief review and a few points only will be mentioned.

The incorporation of 5-amino[5-^{13}C]laevulinic acid, via porphobilinogen (31) into protoporphyrin IX (32) gives the indicated labelling pattern, using an enzyme preparation either from duck's blood or *Euglena gracilis*[50]. The

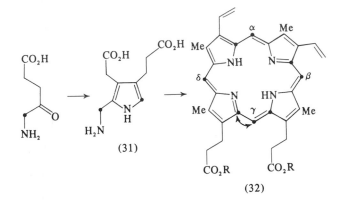

(31)

(32)

incorporation of four intact porphobilinogen units is indicated by C–C couplings. In particular, the coupling ($^1J_{CC} = 72$ Hz) indicated by the double headed arrow in (32) indicates the specific rearrangement of one unit of (31). This study underlines the importance of ^{13}C studies for C—C bond formation, and because the experiment was carried out with a diluted ^{13}C label in the precursor, only C–C coupling within a single incorporated unit of (31) is observable in (32). The assignment of the ^{13}C spectrum of protoporphyrin IX (32) was based upon the synthesis of (32) specifically labelled at selected bridge positions α, β, γ and δ in (32).

A study of vitamin B_{12} biosynthesis[51] using *Propionibacterium shermanii* shows that the indicated ^{13}C labelled precursors are incorporated to give the labelling pattern shown, determined by ^{13}C n.m.r., for isolated cobinamide (34). The incorporation of methionine (33) is of particular interest[52] since the

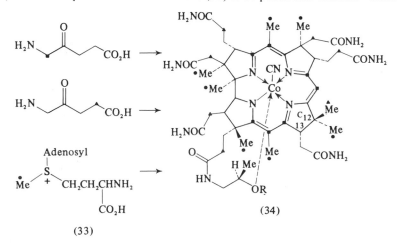

(33)

(34)

[13]C spectrum of isolated cobinamide shows that seven CH_3 groups with shifts in the range 20–27 p.p.m. are labelled, but in the ^{13}C spectrum of neocobinamide (35), which is epimerised at C-13 [see (33) and (35)], one labelled CH_3 is specifically shifted downfield from 23.8 p.p.m. in (34) to 35.5 p.p.m. in (35), which is consistent with the removal of the γ-shielding effect of the C-13 side chain. The labelled methyl group at C-12 in (34) therefore has the α-stereochemistry as shown. This result was obtained in a different manner[53]

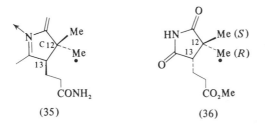

(35)   (36)

by examining the ^{13}C satellites in the ^{1}H spectrum of the degradation product (36), derived from the cobinamide ring C. The CH_3 resonance at τ 8.75, assigned as the *pro-R* methyl group of (36), showed enhanced ^{13}C satellites, indicating isotopic labelling.

Other examples of biosynthetic studies in which C–C coupling has been used as a probe for bond formation include avenaciolide (37), in which $^1J_{CC} = 75$ Hz was observed[54] for C-11 and C-15, using [2-^{13}C]acetate as the precursor, and dihydrolatumicidin (38). The latter study[55] used (i) [13]CH_3-[13]CO_2H as a precursor to indicate the incorporation of intact acetate units on the basis of the preservation of $^1J_{CC}$ (intermolecular coupling disappears owing to dilution with unlabelled material), and (ii) a 1:1 ratio of [13]CH_3CO_2H and CH_3[13]CO_2H to provide information about bond formation between

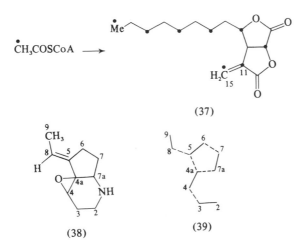

$\overset{\bullet}{C}H_3COSCoA \longrightarrow$

(37)

(38)   (39)

different acetate units since only the type CH_3[13]CO—[13]CH_2COR gives observable $^1J_{CC}$ in the product. These points were illustrated by the ^{13}C spectrum of the metabolite (38), obtained using conditions (i) [C-2–C-3, C-4–C-4a, C-5–C-6, C-7–C-7a and C-8–C-9 coupling observable] and (ii) [C-3–C-4, C-7a–C-4a, C–5-C-8 and C-6–C-7 coupling observable], which indicate acetate coupling as indicated in (39). This method may be used in principle[55] for structure determination of natural products, provided that ^{13}C labels can be introduced with adequate efficiency. A similar technique with doubly labelled acetate was used[56] to show that mollisin (40) is formed by acetate coupling, as indicated in (41).

The lack of quantitative information on the relative amounts of ^{13}C at various labelled positions is a distinct disadvantage of the ^{13}C n.m.r. method,

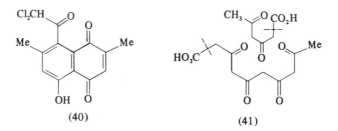

(40)   (41)

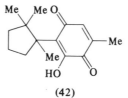

(42)

but the problem may be overcome by the use of the relaxation reagent Cr(acac)$_3$. In a study of the biosynthesis of helicobasidin (42) using this method[57] it was possible to obtain quantitative information about the extent of ^{13}C labelling.

^{13}C Studies of reaction mechanism are still relatively rare, but a study[58] of the generation and rearrangement of phenylcarbene used the ^{13}C labelled sodium salt of benzaldehyde tosylhydrazone (43); the resulting fulvenallene

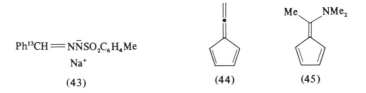

Ph^{13}CH=N$\bar{\text{N}}$SO$_2$C$_6$H$_4$Me
Na$^+$

(43)   (44)   (45)

(44) was trapped by reaction with dimethylamine to give the fulvene (45). The ^{13}C n.m.r. spectrum of (45) indicated a uniform distribution of ^{13}C over every carbon atom of (44).

3.3 SHIFT REAGENTS

3.3.1 Lanthanide reagents

3.3.1.1   Theory

Lanthanide shift reagents became widely used in n.m.r. spectroscopy following the recognition, in 1970, of large lanthanide-induced shifts (LIS)* in the n.m.r. spectra of Lewis bases on the addition of reagents such as Eu(dpm)$_3$ (46). The use of lanthanide reagents has been widely reported during the period under review, a number of reviews have appeared[2-5], and a large number of shift reagents are now commercially available. In particular, the lanthanide complexes Ln(dpm)$_3$ (46) (Ln = lanthanide, most commonly Eu

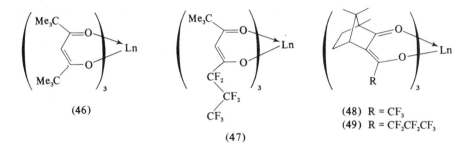

(46)   (47)   (48) R = CF$_3$
(49) R = CF$_2$CF$_2$CF$_3$

and Pr), Ln(fod)$_3$ (47) and the chiral shift reagents (48) and (49)[59-62], based upon camphor and also known as Optishift I and Optishift II, respectively, are commonly used in structural work in organic chemistry.

Shift reagents have largely been used in organic n.m.r. in a qualitative manner for the determination of structure, configuration and conformation in two ways. These are either (i) a rough correlation of the induced shifts with the expected site of complex formation and approximate lanthanide–nuclear distances, or (ii) simply expansion of the observed n.m.r. spectrum to enable coupling constants to be more readily extracted. The chiral reagents find use in the determination of enantiomeric purity and also in potentially more subtle, but qualitative, applications for differentiation of the signals of enantiotopic protons or groups. In addition, lanthanides may be used to locate the position of deuterium in organic compounds and as an aid in the assignment of ^{13}C spectra. These qualitative uses do not generally require great experimental or theoretical rigour, but the quantitative use of lanthanide shifts for the determination of the conformation of organic compounds in solution has excited a certain amount of controversy, and it has become clear that a considerable degree of caution is required. Some aspects of this are discussed in the following section.

The initial theory for the origin of pseudo-contact shifts by the formation of paramagnetic lanthanide–substrate complexes has been revised[63]. For axially

---

* In this section the following symbols will be used: L for lanthanide shift reagent, S for substrate, LIS for lanthanide induced shifts.

symmetrical complexes, provided that certain conditions are obeyed, the relative induced shift of the ith nucleus, ΔS_i, at constant temperature will obey the Robertson–McConnell equation:

$$\Delta S_i = K(3\cos^2\theta_i - 1)/r_i^3 \qquad (3.1)$$

where r_i is distance from the lanthanide atom to the ith nucleus and θ_i is the angle between the lanthanide–ith nucleus vector and the magnetic axis of the complex. The implications of the theory have been discussed[64] in terms of a general strategy for the quantitative use of lanthanide reagents for the determination of conformation and the following rules have been proposed.

(i) Lanthanide reagents (L) giving both high- and low-field shifts should be used. Equation (3.1) should apply, with various values of the constant term K, for all lanthanides so that the ratio of shifts at different sites should be the same for all L. If this condition is fulfilled, then the shifts do have a pseudo-contact origin and the anisotropy of the susceptibility does have axial symmetry as required.

(ii) Diamagnetic corrections should be made using La^{3+} and Lu^{3+} as the complexing lanthanide reagents. On the basis of published data[65,66], these corrections are likely to be relatively small.

(iii) Equation (3.1) applies only for 1:1 complexes and the stoichiometry of the complex should be established by following the change in the induced shift with the concentration of the lanthanide.

Several of these points have been independently considered by other workers. The crystal structures of a number of lanthanide complexes of the type $Ln(dpm)_3S$ indicate that axial symmetry is absent[66-69]. The substrates in these studies have generally been small molecules, and it has been shown[70] that even in the absence of axial symmetry equation (3.1) will still be valid for definable conditions of rapid internal rotation. An interesting paper[71] related to this problem demonstrates a relatively large barrier ($\Delta G^{\ddagger} \gtrsim 7$ kcal mol^{-1}) for rotation in the complex (50), but this compound is related to systems for

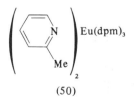

(50)

which equation (3.1) has been shown not always to apply. A study[72] of rigid bicyclic alcohols, based upon Pr- and Eu-induced ^{13}C and ^1H shifts, was consistent with a principal magnetic axis lying close to the Ln—O bond, as required for equation (3.1).

A large number of papers[73-80] have reported on the contact contribution to LIS, generally on the basis of relative shifts for a number of different lanthanide reagents and a single substrate. In many cases the substrates have been aromatic amines, such as quinoline and pyridine, and it seems likely[79] that the most popular reagent, $Eu(fod)_3$, gives rise to the largest contact contributions. Not unexpectedly, ^{13}C shifts are more susceptible to contact contributions than ^1H or ^{19}F shifts[77-80], but the contact shifts can be minimised by the use of the correct lanthanide and it appears that Eu is the worst in this respect and Yb[80] the best. Other advantages of Yb include downfield shifts, which tend to spread the spectrum more than upfield shifts, which are *ca.* 300% larger than those induced by Eu, and little line broadening. In an independent examination[81] of the merits of various lanthanides it was again concluded that Yb is the best for downfield shifts and Dy is the best for upfield shifts. However, most reported studies are still based on Eu, and $Eu(fod)_3$ is generally preferred to $Eu(dpm)_3$ owing to its greater solubility in $CDCl_3$ and CCl_4 [82].

Most studies of geometry have been based on the magnitude of pseudo-contact shifts, although the importance of additional studies using a reagent giving minimal shifts and a maximum relaxation effect has been stressed[64,80,81,83]. The relaxation effect of the *i*th nucleus, as evidenced by increased linewidths, shows a $1/r_i^6$ dependence and $Gd(fod)_3$ has been suggested[83] as the ideal relaxation reagent.

Other paramagnetic metal chelates may give rise to contact shifts and relaxation effects in organic substrates. The relaxation effects show a $1/r_i^6$ dependence and paramagnetic complexes of a number of amino acids have been used in conformational studies based solely upon these effects[84]. Relaxation reagents cause spin decoupling[85] if the rate of relaxation of one of the coupled spins becomes sufficiently fast. Thus the addition of a low concentration (0.0001 M) of $Gd(fod)_3$ to a 0.2 M solution of pyridine in $CDCl_3$ containing $Eu(fod)_3$ causes rapid exchange of H-2 of pyridine between the $|a\rangle$ and $|\beta\rangle$ spin states; this results in broadening of the H-2 resonance and loss of H-2–H-3 coupling.

The methods used for relating molecular geometry to relative values of contact shifts have in most cases been related to equation (3.1). In addition to the tests outlined above it is also necessary to check the stoichiometry of complex formation, and this is done by following changes in the induced shift with changes in lanthanide, L, or substrate, S, concentrations. Several methods have been described for carrying out this test[86-91] and a satisfactory technique is to keep [S] constant and change [L] by adding a stock solution of S to an initial solution of S and L. This procedure minimises handling of the moisture-sensitive lanthanide reagent[89]. On the basis of this test many substrates are shown to form complexes of the 1:1 (LS) type, but quite a large number of 1:2 (LS_2) complexes are also formed, particularly when $Eu(fod)_3$ rather than $Eu(dpm)_3$ is used.

Convincing evidence for the formation of LS_2 complexes has been obtained by n.m.r. studies at low temperatures[91-93], when the equilibrium between reagents and complex becomes, in some cases, slow on the n.m.r. timescale. Again it is evident that $Eu(fod)_3$ with sterically unhindered substrates has a tendency to form LS_2 complexes.

It is also clear that there is a possibility of a change in conformation, or conformational equilibria, on the formation of the complex LS, and an effect of this sort might be detected by changes in coupling constants on complex formation[94-98]. For rigid molecules, with a single conformation, e.g. camphor

(51), small changes in coupling constants are detectable for protons close to the site of complex formation[94,95]. These changes are generally not large, compared with the errors that are inherent in the treatment of vicinal coupling

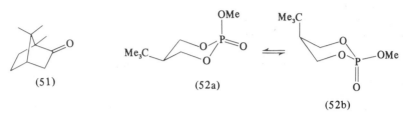

(51)   (52a)

(52b)

by the Karplus relationship. For systems involving an equilibrium between two or more conformations, the equilibrium may be changed on complex formation with resulting changes in vicinal coupling, e.g. for the equilibrium (52a) ⇌ (52b)[97].

The actual computational technique for obtaining molecular geometries in suitable cases from LIS data has been described in outline by a number of different groups[64,99-104]. All use the basic equation (3.1) and generally the best reasonable position for the metal atom is found by fitting some or all of the data. This search may be assisted by the additional use[64] of a relaxation reagent together with the line-broadening dependence upon $1/r^6$. The significance of the fit of LIS data to calculated geometry has been discussed[101] in terms of an agreement factor R. R is evaluated by equation (3.2), where $(\Delta H/H)_{\sigma_i}$ is the observed relative shift for the ith nucleus $(\Delta H/H)_{c_i}$ is the calculated relative shift for the ith nucleus, and w_i are appropriate weighting factors.

$$R = \left[\sum_i \frac{[(\Delta H/H)_{\sigma_i} - (\Delta H/H)_{c_i}]^2 w_i}{(\Delta H/H)_{\sigma_i}^2 w_i} \right]^{\frac{1}{2}} \qquad (3.2)$$

In this particular approach applied to complex formation at oxygen[101], values of R are calculated for movement of the lanthanide atom over the surface of a sphere of suitable radius with the oxygen atom at the centre. R can then be mapped in terms of the two angles which locate the lanthanide on the surface of the sphere and hence the best lanthanide position found. Values of R as low as 0.03–0.09 are obtained and in all cases the computed position of the lanthanide is stereochemically sensible, the O—Ln distance is not critical and atomic coordinates need only be accurate to ca. 0.1 Å. The application of this technique to problems involving configuration and conformation indicates that, with care, alternative possibilities can be assessed and decisions made in many cases at high confidence levels[102]. For example, application of LIS data to isoborneol (53) gives an R value of 0.05, but the same data give an R value of 0.445 when used for the borneol (54) structure. The syn-fused cyclobutane derivative (55) gives an R value of 0.092 for a planar four-membered ring and 0.119 for a non-planar ring, which hardly permits a decision between these alternatives, but the use of the anti structure [cf. (55)] for the same LIS data gives $R = 0.341$, so that clearly the syn stereochemistry is correct.

Although most workers have used computers to determine structures from

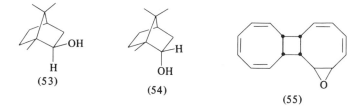

(53) (54) (55)

LIS, a graphical method has also been described[104]. The dipolar field for an axially symmetrical lanthanide complex is mapped on the same scale as a standard Dreiding model; the model may then be placed over the map to obtain relative LIS values and the position of Ln adjusted to give the best fit. It seems likely that for complex molecules the computational approach will continue to be used, together with suitable statistical testing of the calculated geometries. Since the positioning of the lanthanide atom presents problems for most substrates, it is encouraging to learn that, for $Eu(fod)_3$ complexes of nitriles, assumed collinearity of —C≡N—Eu gives good results[105], which stand up to a check using $Yb(dpm)_3$. The introduction of a —C≡N site may therefore prove to be a valuable aid to structure determination using lanthanides.

3.3.1.2 Applications

The configuration and conformation of the epoxide (56), formed by treatment of the unusual natural product (57) with base, were established[106] partly on the basis of the indicated 15% NOE between the substituents on the epoxide ring and partly on the basis of the quantitative use of LIS data. Thus using data obtained from $Yb(dpm)_3$, additional relative values of shifts were obtained, at 220 MHz, for all the protons and methyl groups of (56). The position of the Yb atom was determined on the basis of the $3\text{-}CH_3$ and 4-H substituents on the epoxide ring, then 5-H and 5'-H were located followed by

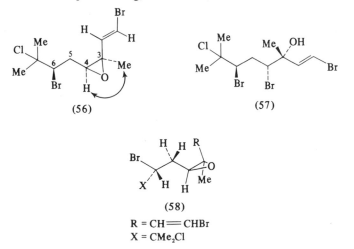

(56) (57)

(58)

R = CH═CHBr
X = CMe_2Cl

6-H with the assistance of coupling constant data. The shifts for C-7–Me$_2$ were then included to give the configuration at C-6; no configuration or conformation other than (58) gave satisfactory agreement with the LIS data. The indicated configuration of the natural product (57) follows from that of (56). The quantitative analysis of the Eu(fod)$_3$ induced shifts of the protons of the adduct (59) of the highly reactive $_\pi 6$ component dibenzo [c,e]tropone (60) with cyclopentadiene established the configuration of the product[107], which indicates *exo* stereochemistry in the $_\pi 4 + _\pi 6$ cycloaddition reaction. The *exo* stereochemistry of (59) was obtained at 99.5% confidence level by the R-factor analysis outlined above, and it is noted that a qualitative assessment of the LIS data had led to the diagnosis of *endo* stereochemistry.

LIS data [Pr(dpm)$_3$] has been used quantitatively, but with a different

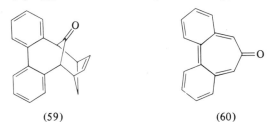

(59)   (60)

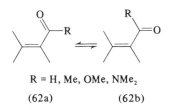

(61)

computational technique, to determine[108] the three-dimensional structure of the antimalarial drug chloroquine (61). In this case the conformational proposal was not subjected to the statistical testing of the R-factor method, and although the stoichiometry of the complex was established as 1:1, the test of the applicability of equation (3.1) using other lanthanide reagents was not carried out.

The application[109] of quantitative LIS evaluation to determine the ratio of the *s-cis* (62a):*s-trans* (62b) conformations of α,β-unsaturated aldehydes,

R = H, Me, OMe, NMe$_2$

(62a)   (62b)

ketones, esters and amides is of interest because details of the computational procedure are given (similar to that of Ref. 101). The position of the lanthanide and calculated LIS values were critically evaluated using the *R*-factor method (or *AF*, Hamilton agreement factors, in the notation of Ref. 109). Where comparisons could be made with data obtained by other methods, agreement was good.

Conformational equilibria in the 2-alkylcyclohexanone–lanthanide chelate complexes [Eu(fod)₃] (63) were examined by a rather different procedure[110]. Characteristic shifts were obtained for equatorial and axial protons at the

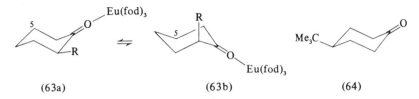

(63a)   (63b)   (64)

various ring positions using 4-t-butylcyclohexanone (64); the shift ratio of a ring proton relative to the summed shifts of both ring protons at that position was then obtained for various lanthanide ratios. In particular, a characteristic ratio was obtained for the equatorial 5-H of (64) and comparison of this ratio with that obtained for the 5-H of the cyclohexanones (63) gives the ratio (63a):(63b). The results obtained in this study agree well with those obtained by other methods for the free ketones.

The quantitative determination of conformation from LIS data remains an area of difficulty and progress has been relatively slow. On the other hand, the qualitative use of LIS has expanded rapidly and it is only possible to refer to a few examples here.

The relative stereochemistry of presqualene alcohol (65) and its synthetic diastereomer (66) were established[111] by comparison of LIS data [Eu(fod)₃]

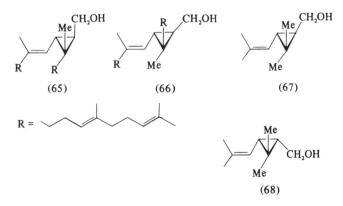

(65)   (66)   (67)

R =

(68)

with data for the analogous chrysanthemyl alcohols (67) and (68). The relative configuration of the tricyclic sesquiterpenoid (69) follows[112] from the observation that the signals of all three methyl substituents are shifted downfield to an equal extent by the addition of Eu(dpm)₃. The structure and

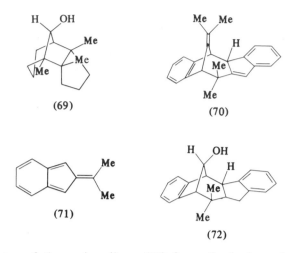

(69)   (70)

(71)   (72)

stereochemistry of the major dimer (70) from the hydrocarbon (71) were established[113] by LIS analysis of the n.m.r. spectrum of the derived alcohol (72). The reactant and product stereochemistry in the 'forbidden' 1,3-sigmatropic rearrangements (73) → (74) were established[114] by a combination of LIS shifts and coupling constant analyses for the alcohol (R = H). The high degree of confidence possible from assignments of this type will be very valuable for the investigation of stereoselectivity in organic reactions.

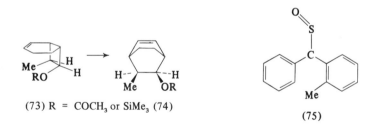

(73) R = COCH₃ or SiMe₃ (74)

(75)

The range of compounds giving complexes and LIS with lanthanide reagents has been extended to include *cis*-azo compounds (but not *trans*-azo compounds)[115]; sulphines[116], e.g. (75), for which configuration can be established; and a number of organic cations[117] in CDCl₃ solution.

3.3.2   Shift reagents other than lanthanides

The use of diamagnetic shift reagents, in which shifts are caused by the induced ring currents in the macrocyclic porphyrin system, has now been fully described[118]. The shift reagent is the germanium tetraphenylporphyrin system (76) and the attached ligand X may be derived from a phenol, an alcohol or a Grignard reagent (with an appropriate ligand Y). The shifts in these systems are to high field, as expected for the diatropic porphyrin system; similar high-field shifts are also observed[119] for the protons of the group R in the

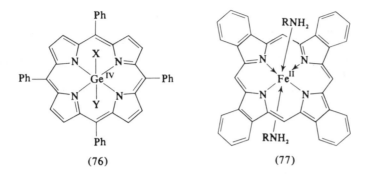

(76)　　　　　　　　(77)

diamagnetic iron phthalocyanine complexes (77). The Co[III] complex of tetraphenylporphyrin has also been suggested[120] as a shift reagent for amines. The tetraphenylborate anion behaves[121] as a shift reagent for sulphonium compounds, and signals in the sulphonium cation are shifted to high field relative to those in other sulphonium salts.

3.3.3 Chiral lanthanide shift reagents, chiral solvents and related topics

The chiral lanthanide shift reagents based upon camphor derivatives were referred to briefly in Section 3.2.1.1. Recent work includes an improved method[122] for the synthesis of the lanthanide complexes of trifluoroacetyl-camphor (78), and a study[123] of the lanthanide complexes of pivaloyl camphor (79).

The trifluoromethyl-substituted reagent (78) has been used[124] to study the optical purity of the β-hydroxy esters (80), which show signal separation corresponding to the two enantiomers at an L/S ratio of *ca.* 0.4. The optical

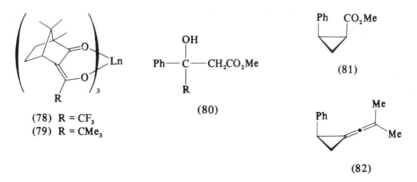

(78) R = CF$_3$
(79) R = CMe$_3$

(80)

(81)

(82)

purity of the cyclopropane derivative (81) was also established[125] using the shift reagent (78), in a study designed to establish the configuration of the related allene (82). The potentially convenient synthesis of chiral alcohols RCHDOH using the chiral lithium aluminium deuteride derivative (83) gives products with optical purities of 20–66%; the estimation of the optical purities of these products would be difficult using optical methods, but the reagent (78) in an L/S ratio of 0.5–0.6 proves satisfactory[126].

The original techniques for the determination of optical purity by n.m.r. methods, using derivatives prepared with chiral reagents, has been extended by the use of lanthanide reagents. Thus the optical purity of α-deuteriated primary alcohols can be determined using the camphanic acid esters (84; H_R or $H_S = D$). The diastereotopic protons of the ester (84) are anisochronous, and the chemical shift difference between them is enhanced by the addition of Eu(dpm)$_3$ in CCl$_4$ solution[127]. In all the cases studied the *pro-R* hydrogen [see (84)] in the shifted spectrum is at higher field than the *pro-S* hydrogen, suggesting that the method may be useful for the assignment of absolute

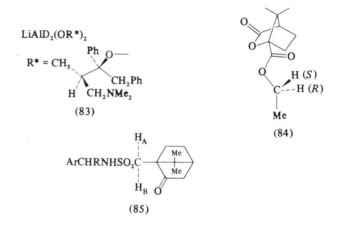

(83)

(84)

(85)

configuration to the deuteriated alcohols, in addition to the estimation of optical purity. The method may also be applicable to α-deuteriated primary amines.

The optical purity and absolute configuration of amines of the type $ArCHRNH_2$ may be determined[128] using the *d*-camphor-10-sulphonamido derivatives (85), since the indicated diastereotopic methylene protons give different chemical shift differences ($\delta_A - \delta_B$) for the two diastereomeric

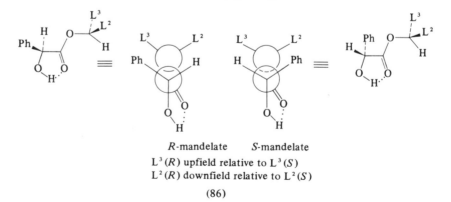

R-mandelate   *S*-mandelate

L^3 (*R*) upfield relative to L^3 (*S*)
L^2 (*R*) downfield relative to L^2 (*S*)

(86)

amides and it is consistently found that $(\delta_A - \delta_B)$ for the (R)-amine is greater than $(\delta_A - \delta_B)$ for the (S)-amine. The absolute configurations of alcohols, L^2L^3CHOH, and amines, $L^2L^3CHNH_2$ (L^2 and L^3 refer to a variety of proton-containing substituents), may well be derivable[129] on the basis of empirically derived correlations of configuration and relative chemical shifts of the group L^2 and L^3, using mandelate, O-methylmandelate, atrolactate and a-methoxy-a-trifluoromethylphenylacetate esters. The method used is based upon, for example, the relative shifts of L^3 and L^2 in the esters derived from R- and S-mandelic acid and similar arguments are proposed for the other esters [see (86) for an explanation].

It has also been found[130] that the salts of chiral bases, such as (87), with enantiomerically pure (S)-a-methoxy-a-trifluoromethylphenylacetic acid (88)

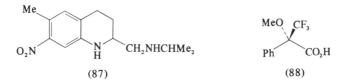

(87)   (88)

give chemical shift differences in $CDCl_3$ associated with R-base–S-acid and S-base–S-acid ion pairs.

Either chiral solvents or chiral shift reagents may be used to distinguish between *meso*- and *d*- or *l*-diastereomers[131,132], and they will also make enantiotopic groups diastereotopic and potentially anisochronous[133]. These principles have all been illustrated for suitable compounds. For example, the *meso*-aminoester (89) shows two C—H proton signals using the chiral alcohol (90) as solvent, whereas the enantiomerically related alcohols (91) and

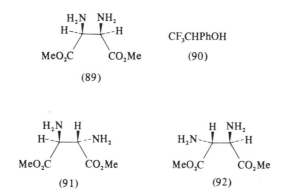

(92) show different sets of signals in the chiral medium, but for each enantiomer only a single C—H signal is observable[132]. Similarly, dieldrin (93) and its metabolite (94) both have enantiotopic pairs of H atoms at C-2 and C-7, C-3 and C-6, and C-4 and C-5 which become diastereotopic and anisochronous in the presence of a chiral shift reagent[134].

The chiral shift reagent (78) may also be used to differentiate enantiomers related as rotational isomers[135], e.g. the two chiral rotational isomers (95a)

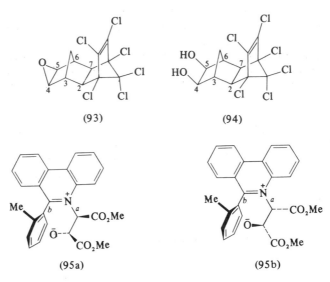

(93)   (94)

(95a)   (95b)

and (95b) each show doubling of each pair of methyl signals on the addition of (78). This observation is consistent with slow rotation (on the n.m.r. timescale) about bonds a and b [see (95)].

3.4 COUPLING CONSTANTS

The use of coupling constants, particularly J_{HH}, for the determination of relative configuration and conformation is long established, and is one of the most important methods available. This review will therefore be restricted to a few recent examples.

The application of the Karplus equation to vicinal H–H coupling constants ($^3J_{HH}$) has been refined recently by using constants in the equation (3.3) based on structures determined by crystallographic methods[136]. The equation was then applied to the conformational analysis of the sugar ring in nucleosides and nucleotides:

$$^3J_{HH} = A \cos^2 \phi_{HH} - B \cos \phi_{HH} + C \tag{3.3}$$

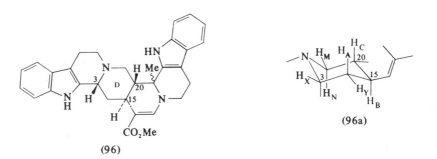

(96)   (96a)

A rather more routine application of $^3J_{HH}$ relationships is exemplified by the determination[137] of the configuration (at C-3, C-15 and C-20) and conformation of ring D of the alkaloid roxburghine B (96). This involved the analysis of an ABCMNXY system [see (96a)] to give the required values of $^3J_{HH}$ (J_{AB}, 12.02; J_{AX}, 4.60; J_{BC}, 11.35; J_{XY}, 2.50 Hz).

$^3J_{HH}$ is also valuable for the conformational analysis of peptides by application of the Karplus relationship to the fragment (97). Experimental values for the constants A, B and C of equation (3.3) agree[138,139] reasonably well with

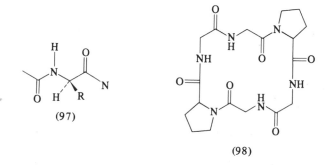

(97)

(98)

theoretical values in the range $\phi = 0–90°$, but less well in the range $\phi = 90–180°$. The application of n.m.r. techniques to the determination of peptide conformation is also illustrated[140] by the interpretation of the 250 MHz ^1H and 25.16 MHz ^{13}C spectra of cyclo (Gly-L-Pro-Gly)$_2$ (98).

Long-range proton–proton coupling continues to receive attention, although extensively documented. Thus pathways for observable $^4J_{HH}$ other than the well known W pathway have been found[141] in norbornene derivatives. The conformational dependence of $^5J_{HH}$ (homo-allylic coupling) in systems related to (99) has been noted and analysed[142] in empirical and theoretical

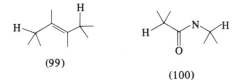

(99)

(100)

terms. $^5J_{HH}$ is also observable for the unit (100) of *trans*-peptide linkages, and although it is often too small for resolution it can be detected by double-resonance techniques[143].

Fluorine–proton coupling ($^3J_{HH}$) may also be analysed[144] on the basis of a Karplus-like relationship (3.3) in a rather similar manner to $^3J_{HH}$, as illustrated by the determination of the conformations of *cis*- and *trans*-4-fluoro-L-proline (101). Long-range proton–fluorine coupling may involve interactions through space, and it has often been observed in sterically crowded situations. The value of $^7J_{HF}$ (7 Hz) in *trans*-1,1'-difluorotetrabenzopentafulvalene (102) is unusually large[145], and it reflects the close contact of the indicated hydrogen and fluorine atoms in this highly strained ethylene derivative.

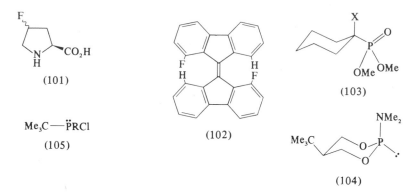

(101)

(102)

(103)

Me₃C—P̈RCl

(105)

(104)

A number of observations of $^3J_{PH}$ indicate a Karplus-like relationship in phosphine oxides[146] and phosphonates, e.g. (103)[147], but for tricovalent phosphorus derivatives, e.g. (104)[148] and (105)[149], $^3J_{PH}$ depends upon both the torsional angle and the orientation of the lone-pair electrons on the phosphorus atom.

The stereochemical dependence of $^1J_{CH}$ has been recognised and this may be useful in assigning configuration[150] to the anomeric (C-1) position of carbohydrates (1-H axial, $^1J_{CH} = 158$–162 Hz; 1-H equatorial, $^1J_{CH} = 169$–171 Hz). In ring systems such as oxaziridines (106)[151], both $^1J_{CH}$ and $^2J_{NH}$ show configurational dependence [for (106; R¹ and R² cis), $^1J_{CH} = 181.5$–185.6 Hz; for (106; R¹ and R² trans), $^1J_{CH} = 173$–180 Hz].

Long-range carbon–fluorine coupling is observable in systems analogous to those in which 'through space' long-range J_{HF} and J_{FF} are observed, e.g. the phenanthrene (107) shows[152] $J_{CF} = 24.2$ Hz between the 4-fluoro substituent and the carbon of the 5-methyl group. Long range C–F coupling is also

(106)

(107)

(108)

observable in aromatic fluoro-compounds[153] and, for example, $^6J_{CF}$ in 2-fluoronaphthalene (108) involving C-6 is 2.8 Hz; coupling of this type is apparently related to a variation in π-bond order between the interacting spins.

A number of papers have appeared relating J_{PC} to structure. In particular the stereochemical dependence of $^2J_{PC}$ has been noted in a number of investigations, and it may be applied for configurational assignments for the substituents on phosphorus in cyclic phosphonates (109)[154] and cyclic phosphines (110)–(112)[155-157]. $^2J_{PC}$ also shows dihedral angle dependence in

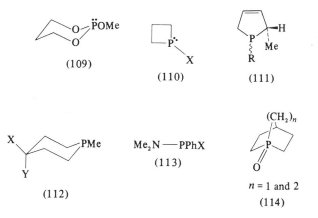

(109) (110) (111)

(112) (113) (114)

$n = 1$ and 2

acyclic tricovalent phosphorus derivatives, and differences are particularly striking for the two methyl carbons of the dimethylaminophosphines (113)[158].

On the basis of the values of $^3J_{PC}$ observable in bicyclic phosphine oxides (114), and other cyclic phosphine oxides, it has been suggested[159] that $^3J_{PC}$ shows a Karplus-like relationship similar to those well established for other vicinal coupling constants. In an analysis of the ^{13}C and ^1H n.m.r. of nucleotides and polynucleotides it has been noted that variations in $^3J_{CP}$ may be useful for conformational analysis[160]; however, in a detailed study[161] of the conformation of 3',5'- and 2',3'-cyclic nucleotides it was shown that $^3J_{HH}$ and $^3J_{PC}$ are less satisfactory for conformational studies than $^3J_{PH}$, owing to the greater sensitivity of the P–H coupling constants to changes in dihedral angle.

The measurement of $^{13}C-^{13}C$ coupling for samples containing only natural abundance ^{13}C represents a rigorous test of spectrometer sensitivity. Two groups[162,163] have reported the observation of C–C coupling under these conditions using pure liquids with ^1H decoupling, and comparisons have been made with analogous C–H coupling. A number of other studies of J_{CC} using ^{13}C enriched samples have led to the accumulation of a useful amount of information on these coupling constants which have obvious application in structural studies, as well as in biosynthetic and mechanistic investigations (see Section 3.2.3.3).

3.5 STRUCTURAL STUDIES USING T_1 AND RELATED EFFECTS

3.5.1 T_1 measurements

The primary reason for the introduction of FT n.m.r. is to obtain increased sensitivity, but, in addition to sensitivity enhancement, FT spectrometers are particularly well suited to the measurement of relaxation times, particularly the spin–lattice relaxation time, T_1, and it is becoming apparent that T_1 measurement may be used in structural studies[1,164].

T_1 Studies using ^1H n.m.r. have not yet found much application in structural examination, but it has been shown that the measurement of T_1 is a valuable complement to shift measurements in investigations based upon the use of lanthanide reagents (Section 3.3.1.1). In a study of carbohydrate derivatives[165] it has been noted that the axial (β) anomeric proton has a shorter T_1 than the corresponding equatorial (a) proton, indicating that there may be correlations of T_1 with configuration and conformation.

T_1 Measurements for ^{13}C have been much more extensively examined, and general correlations of T_1 with structure are beginning to emerge. Simple long-chain molecules[166,167], aromatic compounds[168], and, more relevant for general structure determination, complex polycyclic compounds[169,170] have been studied. T_1 Measurements have also found application for studies of biopolymers[171], but the potential for structural studies of organic compounds lies in the correlation of ^{13}C T_1 values with structural features.

For a protonated carbon[167] the ^{13}C spin–lattice relaxation mechanism is frequently dominated by dipole–dipole interactions with the attached protons; for small molecules in liquids of low viscosity, T_1 is given by:

$$1/T_1 = \hbar^2 \gamma_C^2 \gamma_H^2 \sum_i r_{CH_i}^{-6} \tau_C \qquad (3.4)$$

where γ_H and γ_C are the gyromagnetic ratios for ^1H and ^{13}C, r_{CH} is the C—H distance for all protons contributing to relaxation, and τ_C is the correlation time for molecular rotational reorientations. The values of T_1 for all the carbons of neat decan-1-ol at 42°C are given in (115)*, together with calculated values of τ_C. The values of τ_C given in (115) show that molecular

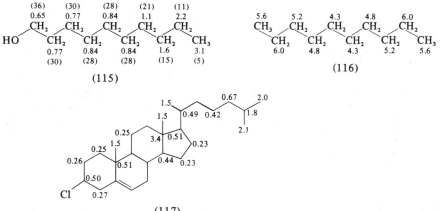

motion becomes more restricted and T_1 shorter as the hydroxyl end of the chain is approached, which is consistent with intermolecular hydrogen bonding. For straight-chain hydrocarbons, τ_C values tend to be dominated by overall molecular reorientation and give shorter values (and longer T_1) than for compounds such as decanol with functional groups that give rise to intermolecular interactions[166]. Decane, for example, gives rather long T_1 values

---

* In formulae (115)–(119) the figures indicate T_1 values in seconds; in (115) the figures in brackets refer to calculated values of τ_c in ps.

that exhibit a minimum near the centre of the chain where motion is more restricted [see (116)]. For more complex polycyclic molecules, such as cholesteryl chloride (117)[169,170], side-chain carbons show rather longer relaxation times than the protonated carbons of the rather rigid polycyclic skeleton. In particular, the side-chain carbons show longer relaxation times towards the end of the chain. Unprotonated carbons can also be picked out because of their significantly longer relaxation times. The data in (117) indicate how T_1 can be related to structural environment for ^{13}C, and T_1 measurement is now beginning to prove an aid in making ^{13}C spectral assignments.

The ^{13}C n.m.r. spectra of codeine (118) and brucine (119) could be partly assigned[172] by the usual decoupling techniques combined with empirical

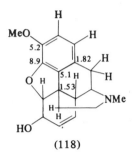

(118)

chemical shift estimates, but the quaternary carbons were difficult to assign. On the basis of equation (3.4), assuming that these carbons are relaxed by an intramolecular dipolar mechanism, the relaxation times should be related to the number of protons bound to a-carbon atoms. The measured values of T_1 are shown in (118) and (119) together with the relevant hydrogen atoms which were noted in making the assignments[172]. In these cases the carbons are all ring carbon atoms and similar values of τ_C are therefore expected [cf. (117)].

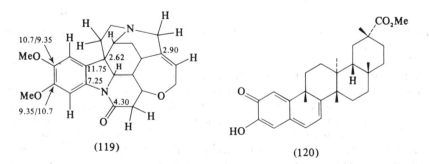

(119)   (120)

In another very instructive example of this technique[173] the carbon atoms of pristimerin (120) and related triterpenes were picked out of the very complex ^{13}C n.m.r. spectra on the basis of their differing relaxation times. In particular, carbons were readily identified as belonging to one of three types: quaternary and carbonyl, methylene and methine, and methyl.

True values for relaxation times can only be obtained for completely deoxygenated solutions; it has recently been shown that the usual freeze–thaw sequence for degassing is inadequate, and a superior chemical technique has been proposed[174]. In simple molecules, such as substituted aromatic compounds, relaxation mechanisms other than that described by equation (3.4) may operate[168]. These mechanisms have been discussed in detail and illustrated by examples. In spite of these potential complications it seems likely that T_1 measurements will play a useful part in structural studies using ^{13}C n.m.r.

3.5.2   Nuclear overhauser effect

The theory and application of the nuclear Overhauser effect (NOE) has been reviewed in detail[175]. A theory, originally developed for rigid molecules, has been extended to the more general case of non-rigid molecules[176]. The essential results are that when the rate of exchange between conformations (k) is slow compared with the relaxation rate of the spins (R_d), the observed NOE is an ensemble average of the effects characteristic of each conformation. On the other hand, if $k \gg R_d$ the ensemble average spin interactions must be used to compute the NOE. This theory has been applied to the NOEs observable in the ^1H spectrum of 2′,3′-isopropylidene-inosine (121) and -neuridine (122). A

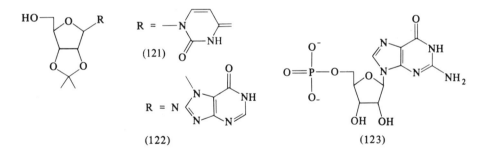

quantitative study of guanosine monophosphate (123) has also been reported[177].

The bridged biphenyls (124) were also studied by quantitative NOE, which required the extension of the theory to three-spin systems[178]. The majority of studies using the NOE for the study of organic compounds, however, continue to be qualitative and only a few representative examples will be cited.

The position of deuteration in the alkaloid derivative (125) was shown by the 16% NOE between C-9—H and C-10—OCH₃, and this conclusion was confirmed using Eu(fod)₃ induced shifts[179]. This example is worthy of note since it is relevant to orientational problems in substituted aromatic compounds that may be difficult to solve if informative coupling information cannot be obtained.

The stereoselectivity of base-catalysed H/D exchange in sulphur compounds has occupied considerable attention in recent years. On the basis of the NOE observable in (126b) (obtained from the tetradeuterio salt by D/H exchange), the α-protons in the sulphonium salt (126a) which undergo most rapid H/D exchange are *cis*- to the S—Me substituent[180]. The doubly bridged helicene (127) was obtained in two isomeric forms by photocyclisation of (128); these isomers could be *meso* or *dl*, depending on the relative helicities of the two

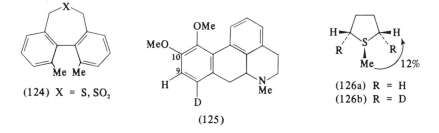

(124) X = S, SO₂

(125)

(126a) R = H
(126b) R = D

helicene systems[181]. The major isomer was shown to have the *dl*-stereo-chemistry shown in (127) by the observation of a 25% NOE between 2-H (≡ 14-H) and 1-H (≡ 13-H). These pairs of protons are close to one another in the *dl*-isomer but remote in the *meso*-isomer.

The NOEs of sesquiterpenes with 10-membered rings have proved extremely useful for the assignment of configuration and conformation, but caution is required when more than one stereoisomer is present. Thus the sesquiterpene (129) at low temperatures adopts two conformations[182], giving rise to two sets of n.m.r. signals in the ratio 92:8. The NOEs indicated in (129) are consistent with the conformation shown and in particular the NOEs

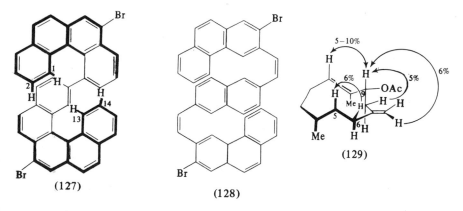

(127)

(128)

(129)

1-H → 9β-H and 5-H → 6β-H are diagnostic. At +30°C the second conformation is in rapid equilibrium with the first (129) and anomalous NOEs are observable owing to contributions from this minor conformation (*cf.* Ref. 176). The 10-membered ring of the sesquiterpene lactone costunolide (130) has been shown to have the indicated conformation on the basis of NOEs readily observable in the spectrum spread out by the addition of Eu(fod)₃ [183]. This technique has advantages associated with the spread out spectrum, but there is some reduction in NOE intensities owing to relaxation time decreases caused by the paramagnetic shift reagent, particularly for protons near the lanthanide binding site.

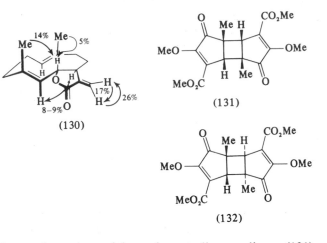

(130)

(131)

(132)

The relative configurations of the cyclopentadienone dimers (131) and (132) cannot readily be determined[184] on the basis of NOEs, since, although at high observing field strengths the NOEs appear diagnostic [(131) Me → H, 42%, (132) Me → H, 15%], at lower observing field strengths the NOEs of both (131) and (132) become identical (Me → H, 40%). The similarity of T_1 for the $CO_2C\underline{H}_3$ and $OC\underline{H}_3$ protons in both cases suggests similar values of τ_C [*cf.* equation (3.4)], but the T_1 for the cyclobutane protons of *syn*-configuration (131) (T_1 3.40 s) is shorter than that for the *anti*-configuration (132) (T_1 4.75 s), as expected on the basis of shorter H—H distances in (143). The problems posed by (131) and (132) are similar to those in other structural problems involving dimer formation and ^1H T_1 measurement may find application for these cases. The results also underline the need for caution in the interpretation of NOE results.

3.6   THE STUDY OF CONFORMATIONS AND CONFORMATIONAL CHANGES

The use of the temperature dependence of n.m.r. lineshapes to study the rates of conformational changes has continued to be a popular method for studying conformational energy profiles. The important advances in these studies are (i) the use of higher field spectrometers, (ii) the use of FT spectrometers for ^{13}C studies, and (iii) the use of achiral and chiral shift reagents. These aspects will be covered by the papers selected for review.

Shift reagents such as Eu(fod)$_3$ may be used, for suitable substrates, to increase the separation of signals, observable at low temperatures, due to different conformers of the same substrate. As had already been noted, complexed substrates may not have the same equilibrium constants for conformational equilibria as free substrates, and this effect must also be considered when evaluating the significance of measured energies of activation. Provided that the L/S ratio is reasonably low, and the exchange LS + S' ⇌ S + LS' is fast, the measured activation energy should correspond to a pathway involving the lower energy transition state. The technique has been checked[185] using amides (133) and Eu(fod)$_3$. $\Delta G^{\ddagger}$ obtained by the coalescence temperature method shows a slight dependence on the Eu(fod)$_3$ concentration, but otherwise agreement with literature values is good. As expected, ($\nu_A - \nu_B$) for the two N—Me groups is temperature dependent, but approximately linear in $1/T$ so that the value at the coalescence temperature can be obtained by extrapolation.

Rather similar results were obtained for the urethane (134)[186], although in this case ($\nu_A - \nu_B$) is too small to be measured in the absence of Eu(fod)$_3$.

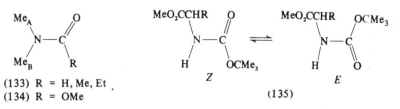

(133) R = H, Me, Et
(134) R = OMe

(135)

The free energy barrier for rotation about the N—CO bond of (134) was estimated to be 15.5 kcal mol^{-1} by extrapolation to zero concentration of shift reagent. The amino-acid derivatives (135) exist as Z- and E-diastereomers; the E-isomers complex more readily with Eu(fod)$_3$ and the proportion of Z- and E-isomers therefore changes with temperature. Making allowances for both this effect and the temperature dependence of chemical shifts, it was possible to estimate values for $\Delta G^{\ddagger}$ ($Z \rightarrow E$) at the appropriate coalescence temperatures.

If chiral shift reagents or chiral solvents are used it is possible to detect conformational changes involving the interconversion of enantiomeric conformations (giving diastereomeric complexes), which are normally only detectable under 'achiral' conditions using diastereotopically related protons in substituents such as CH$_2$X and CXMe$_2$. A number of examples of this type have been noted. The enantiomeric R- and S-conformations of the diazetidinone (136) may be separately detected[188] at low temperatures ($-55°C$) using a CDCl$_3$ solution of (136) containing equimolecular proportions of

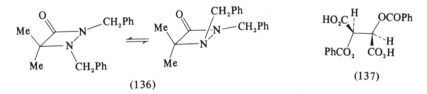

(136)

(137)

(2R,3R)-dibenzoyltartaric acid (137); this leads to a doubling of each of the two methyl signals of (136). The nitrosamine (138) similarly shows separate signals from the R- and S-conformations in (+)-2,2,2-trifluoro-1-phenylethanol, which coalesce to a single set of signals at higher temperatures when rotation about the Ar—N bond is fast on the n.m.r. timescale[188]. The free energy of activation ($\Delta G^{\ddagger}$ 17.1 kcal mol^{-1}) for the interconversion of the enantiomeric conformations in the chiral solvent is similar to that observed in achiral solvents, using the diastereotopic protons of the CH$_2$Ph group as a

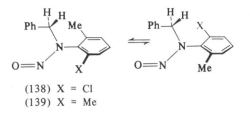

(138) X = Cl
(139) X = Me

probe. The achiral nitrosamine (139) shows different signals[189] from the enantiotopic arylmethyl groups or CH_2Ph protons under conditions of slow rotation about the N—Ar bond (0°C) when they become diastereotopic either in a chiral solvent or in the presence of a chiral lanthanide reagent (Eu Optishift I). These signals coalesce at higher temperatures, and $\Delta G^{\ddagger}$ for rotation about the Ar—N bond may be obtained.

The styrene derivatives (140) give two enantiomeric conformations which are detectable[190] at low temperatures by n.m.r. spectra of solutions containing

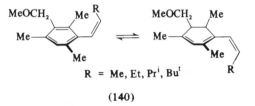

R = Me, Et, Pr^i, Bu^t

(140)

Eu Optishift I. The two OCH_3 signals observable at low temperatures coalesce at high temperatures when rotation about the Ar—C= single bond becomes fast on the n.m.r. timescale.

More conventional methods of studying conformational changes continue to be widely used. The advantages of ^{13}C n.m.r. have been demonstrated for a number of systems in which the proton spectra are either too extensively coupled for use, or the chemical shift differences are too small for detection. For example, the conformational equilibrium (141a) ⇌ (141b) may be studied[191] using ^{13}C n.m.r. at low temperatures, giving, indirectly, a value for

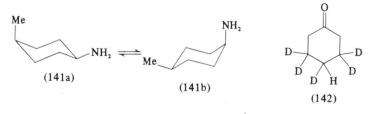

(141a)   (141b)   (142)

ΔG_{NH_2}, although the proportion of the axial conformer (2%) is too low to detect for cyclohexylamine itself.

The barrier to ring inversion for cyclohexanone has been sought for a long time by n.m.r. studies; it has now been measured[192] using 251 MHz n.m.r. and the deuterated cyclohexanone (142). The 4-H of (142) separates into two lines below −184°C ($\Delta G^{\ddagger}$ 4.0 ± 0.1 kcal mol^{-1} for ring inversion). This is the lowest barrier so far measured by the n.m.r. method and the very low temperatures required were maintained by using helium as the cooling gas. The result was in close accord with the inversion barrier predicted by molecular mechanics calculations. Other results of interest for six-membered ring systems include a study[193] of the bis-methylene cyclohexane (143) and the methylene cyclohexanone (144). Full details of the measurement of inversion barriers for a variety of six-membered heterocyclic rings (145) have also been reported[194].

Studies of polycyclic six-membered ring systems include the *cis*-decahydroquinoline system (146)[195], and the complex perhydroanthracene derivative

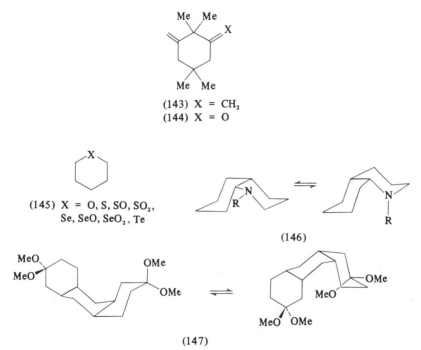

(143) X = CH$_2$
(144) X = O

(145) X = O, S, SO, SO$_2$,
       Se, SeO, SeO$_2$, Te

(146)

(147)

(147)[196]. In both cases evidence was obtained for the indicated conformational equilibrium. The manxine system (148) is also interesting and it has been shown[197] that conformational interconversion of the two enantiomeric conformations (148) is slow on the n.m.r. timescale at −80°C. The racemisation of highly hindered non-planar ring systems analogous to (149) has been studied by classical polarimetric methods, and the study has now been extended[198] by n.m.r. lineshape methods.

Saturated seven-membered ring systems present severe difficulties for conformational analysis and only a few studies have appeared. That of the cycloheptane derivatives (150) and (151) is one of the most conclusive[199] and the evidence for preferred twist-chair conformations [see (150) and (151)] is fully discussed.

Eight-membered ring systems have been extensively investigated. The

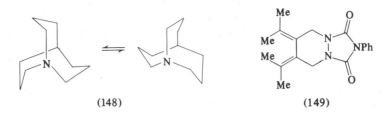

(148)   (149)

unsubstituted cyclo-octane system has been carefully studied by ^{13}C and ^{1}H n.m.r.[200]. The ^{13}C spectrum shows a second signal due to only 0.3% of a minor conformation at low temperatures, which causes broadening of the major signal in the range −20 to −75°C. The 251 MHz ^{1}H spectrum is

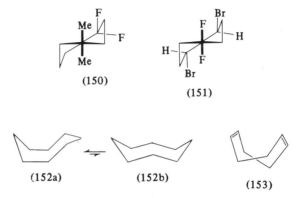

consistent with the presence of two conformational types and slow ring inversion at low temperatures. The major conformation is assigned as a rapidly pseudo-rotating boat-chair conformation (152a) and the minor conformation to the crown family of conformations (152b). The advantages of the complementary use of ^{13}C and ^{1}H n.m.r. are shown very clearly in this and other studies of medium-sized ring systems. The cyclo-octa-1,5-diene system (153) also illustrates this point[201]. The 251 MHz ^{1}H spectrum of the methylene groups splits into two singlets below −168°C which become two doublets below −173°C, corresponding to four different proton environments. The ^{13}C spectrum splits into two signals below −174°C, corresponding to two different methylene groups. These results are interpreted in terms of the twist-boat ground state shown in (153); again the free energies of activation associated with the conformational changes ($\Delta G^{\ddagger}$ 4.4 and 4.9 kcal mol^{-1}) are among the lowest measured by n.m.r. methods.

A number of other eight-membered ring systems, having one or more ring oxygen atoms, have also been examined using ^{1}H n.m.r.[202,203], and in one case ^{1}H and ^{13}C n.m.r.[204]. The dibenzocyclo-octa-1,5-diene system (154) shows ^{1}H signals assignable to chair and boat conformations[205,206] and the free energies of activation for the process chair ⇌ boat and for the inversion of the boat conformation[206] have been measured. The results show a reasonably good correlation with molecular mechanics calculations, and it is of interest that heterocyclic analogues of (154)[205] show similar conformational behaviour. Inversion barriers for the boat-shaped conformation of a suitably substituted benzocyclo-octatetraene (155)[207] ($\Delta G^{\ddagger}$ 13.4 kcal mol⁻¹) and a dibenzocyclo-octatetraene (156)[208] ($\Delta G^{\ddagger}$ 12.3 kcal mol⁻¹) have been measured. In contrast to systems such as (153) and (154), the benzene-annelated cyclo-octatetraenes show rather lower inversion barriers than an analogous cyclo-octatetraene (157) ($\Delta G^{\ddagger}$ 14.8 kcal mol⁻¹).

Ten-membered ring systems present challenging problems for conformational analysis, and in most cases the n.m.r. method does not provide sufficient

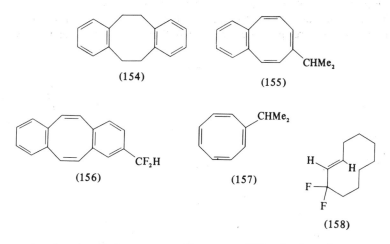

(154)

(155)

(156)

(157)

(158)

information for detailed comment. However, ¹⁹F substituted systems often give simpler n.m.r. results than the unsubstituted hydrocarbons, and although the low temperature ¹H and ¹³C spectra of cyclodecane are not resolved[209], even at −160°C, 1,1-difluorocyclodecane gives an ¹⁹F AB system below −135°C, indicating a low barrier ($\Delta G^{\ddagger}$ 5.7 kcal mol⁻¹) for the ring inversion process[210], but not providing definitive information about the conformation of the 10-membered ring. The low-temperature ¹⁹F n.m.r. spectrum (¹H decoupled) of 3,3-difluoro-trans-cyclodecene (158) is more informative[211]. Five AB systems, corresponding to five of the eight possible conformational types, are observable at −152°C, and these coalesce to a single AB system at −30°C. This corresponds to rapid interconversion of the different conformational types with slow ring inversion, since the latter process involves rotation of the trans double bond through the plane of the ring system. Finally, at +74°C all processes become fast on the n.m.r. timescale and only a singlet ¹⁹F signal is observable.

Cyclodecanone has been studied using the ¹³C n.m.r. spectrum [212]. At −160°C the a, β- and γ-carbons [see (159)] all give 1:1 doublets although the carbonyl carbon and ε-carbon give single resonances. This is consistent with the boat-chair-boat conformation shown (159); the 251 MHz ¹H n.m.r. spectrum is also in accord with this conformation. A study of a number of cycloalkane systems using 63.1 MHz ¹³C n.m.r. provides information concerning the conformations of cyclododecane and cyclotetracosane[209]. The former gives two ¹³C signals in a 2:1 ratio, below −108°C, consistent with the square conformation (160) (D_4 symmetry) having four 'corner' carbons [shown as circles in (160)] and eight non-corner carbons [+ and — in (160)]. Cyclotetracosane shows four types of carbon in a 2:2:2:1 intensity ratio at −132°C, consistent with the conformation shown in (161), in which the four types of carbon are indicated by A, B, C and D. The 12-membered ring system of the hydrocarbon (161a) can exist in two distinct conformational types with C_3 and C_2 symmetry. The temperature dependence of the n.m.r. spectrum is consistent with only the conformation of C_2 symmetry on the basis of a full lineshape analysis[213]. This result is consistent with molecular mechanics calculations even for this highly complex system.

The conformations of cyclophane systems are highly suitable for n.m.r. examination and are often better defined than those of other ring systems.

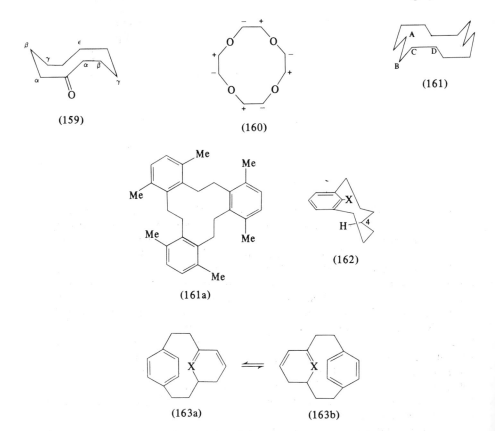

(159)

(160)

(161)

(161a)

(162)

(163a) ⇌ (163b)

[7]Metacyclophane (162; X = H) is concluded[214] to adopt the conformation shown on the basis of the high field signal (δ −1.33), assignable to the indicated 4-H, observable in the n.m.r. spectrum at −73.5°C; the analogous hydrogen in the conformationally rigid bromo derivative (162; X = Br) is observable at δ −1.86. The barrier to rotation of the heterocyclic ring, (163a) ⇌ (163b), in the metaparacyclophane (163; X = N) ($\Delta G^{\ddagger} \sim 11$ kcal mol^{-1}) shows very convincingly[215], in accord with previous studies, that the

nitrogen atom of a pyridine ring has significantly lower steric requirements than the CH group of a benzene ring [$\Delta G^{\ddagger}$ ~20 kcal mol^{-1} for (163; X = CH)]. The metaparacyclophane system has also been used[216] for a very interesting study of kinetic isotope effects; k_D/k_H for (163; X = CD or CH) was determined for the process (163a) $\rightleftharpoons$ (163b), using the very precise double-resonance method described by Forsén and Hoffman. The result, k_D/k_H 1.20 ± 0.04, is the largest conformational kinetic isotope effect yet measured, consistent with the very large steric interaction between X and the para-substituted benzene ring [see (163)] in the transition state for the conformational change (163a) $\rightleftharpoons$ (163b). A similar method has been used[217] to measure the effects of different *para* substituents (relative to X) on the rate of the same process (163a) $\rightleftharpoons$ (163b).

A considerable number of papers have appeared discussing rotational barriers about single bonds, but space permits the inclusion of only three examples. The hindered rotation of a methyl substituent in highly substituted tryptycene systems is of particular interest. The bridgehead methyl substituent in (164) gives an AB$_2$ system at low temperatures[218], corresponding to the conformation shown; the rotational barrier about the C—Me bond is quite substantial ($\Delta G^{\ddagger}$ 10.6 kcal mol^{-1}). A second example, with a rather lower rotational barrier ($\Delta G^{\ddagger}$ 7.2 kcal mol^{-1}) is the quinonoid system (165)[219]. In this case the ^{1}H methyl spectrum for (165; R = CH$_3$) is barely resolved at

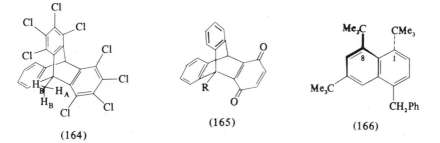

(164)   (165)   (166)

−141°C, but the dideuterio derivative (165; R = CD$_2$H) gives clearly resolved singlet signals, with a 2:1 intensity ratio, at −141°C.

The 1,8-di-t-butylnaphthalene derivative (166) shows non-equivalence of the C-1 and C-8 t-butyl methyl groups at low temperatures, consistent with a low rotational barrier about the Ar—CMe$_3$ bonds ($\Delta G^{\ddagger}$ 6.5 kcal mol^{-1})[220]. The methylene protons of the benzyl substituent are observable as an AB system, even at 195°C, indicating an out-of-plane distortion of the naphthalene system, presumably involving the C-1 and C-8 t-butyl substituents as indicated in (166), with a high energy barrier for the passage of one t-butyl group past the other.

References

1. Gillies, D. G. and Shaw, D. (1972). *Ann. Reports N.M.R. Spectrosc.*, **5A**, 560
2. von Ammon, R. and Fischer, R. D. (1972). *Angew. Chem. Int. Ed. Engl.*, **11**, 675
3. Mayo, B. C. (1973). *Chem. Soc. Rev.*, **2**, 49

4. Cockerill, A. F., Davies, G. L. O., Harden, R. C. and Rackham, D. M. (1973). *Chem. Rev.*, **73**, 553
5. Reuben, J., (1973). *Progr. N.M.R. Spectrosc.*
6. Jesson, J. P., Meakin, P. and Kneissel, G. (1973). *J. Amer. Chem. Soc.*, **95**, 618
7. Feeney, J. and Partington, P. (1973). *J. Chem. Soc. Chem. Commun.*, 611
8. Witanowski, M. and Webb, G. A. (1972). *Ann. Reports N.M.R. Spectrosc.*, **5A**, 395
9. Mavel, G. (1973). *Ann. Reports N.M.R. Spectrosc.*, **5B**, 1
10. Inch, T. D. (1972). *Ann. Reports N.M.R. Spectrosc.*, **5A**, 305
11. Kotowycz, G. and Lemieux, R. U. (1973). *Chem. Rev.*, **73**, 669
12. Eliel, E. L. (1972). *Angew. Chem. Int. Ed. Engl.*, **11**, 739
13. Martin, G. J. and Martin, M. L. (1972). *Progr. N.M.R. Spectrosc.*, **8**, 166
14. Gunther, H. (1972). *Angew. Chem. Int. Ed. Engl.*, **11**, 861
15. Hoffmann, E. G., Stempfle, W., Schroth, G., Weimann, B., Ziegler, E. and Brandt, J. (1972). *Angew. Chem. Int. Ed. Engl.*, **11**, 375
16. Stothers, J. B. (1972). ^{13}C *N.M.R. Spectroscopy* (New York: Academic Press)
17. Levy, G. C. and Nelson, G. L. (1972). ^{13}C *N.M.R. for Organic Chemists* (New York: Wiley-Interscience)
18. Johnson, L. F. and Janowski, W. C. (1972). ^{13}C *N.M.R. Spectra* (New York: Wiley-Interscience)
19. Gansow, O. A., Burke, A. R. and LaMar, G. N. (1972). *J. Chem. Soc. Chem. Commun.*, 456
20. Barcza, S. and Engstrom, N. (1972). *J. Amer. Chem. Soc.*, **94**, 1762
21. Sadler, I. H. (1973). *J. Chem. Soc. Chem. Commun.*, 809
22. Seiter, C. H. A., Feigenson, G. W., Chan, S. I. and Hsu, M. (1972). *J. Amer. Chem. Soc.*, **94**, 2535
23. Birdsall, B., Birdsall, N. J. M. and Feeney, J. (1972). *J. Chem. Soc. Chem. Commun.*, 316
24. Newmark, R. A. and Hill, J. R. (1973). *J. Amer. Chem. Soc.*, **95**, 4435
25. Pines, A. and Ellet, J. D., jr. (1973). *J. Amer. Chem. Soc.*, **95**, 4437
26. Breitmaier, E., Jenny, C., Voelter, W. and Pohl, L. (1973). *Tetrahedron*, **29**, 2485
27. Tulloch, A. P. and Mazurek, M. (1973). *J. Chem. Soc. Chem. Commun.*, 692
28. Colli, H. N., Gold, V. and Pearson, J. E. (1973). *J. Chem. Soc. Chem. Commun.*, 408
29. Stothers, J. B., Tan, C. T., Nickon, A., Huang, F., Sridhar, R. and Weglein, R. (1972). *J. Amer. Chem. Soc.*, **94**, 8581
30. Hunter, D. H., Johnson, A. L., Stothers, J. B., Nickon, A., Lambert, J. L. and Covey, D. F. (1972). *J. Amer. Chem. Soc.*, **94**, 8582
31. Stephenson, L. M., McClure, D. E. and Sysak, P. K. (1973). *J. Amer. Chem. Soc.*, **95**, 7888
32. Jensen, R. J. and Petrakis, L. (1972). *J. Magn. Resonance*, **7**, 105
33. Liebfritz, D. and Roberts, J. D. (1973). *J. Amer. Chem. Soc.*, **95**, 4996
34. Eggert, H. and Djerassi, C. (1973). *J. Amer. Chem. Soc.*, **95**, 3710
35. Jones, A. J. (1972). *Tetrahedron Lett.*, 4351
36. Bhacca, N. S., Wiley, R. A., Fischer, N. H. and Wehrli, F. W. (1973). *J. Chem. Soc. Chem. Commun.*, 614
37. Lukacs, G., Sepulchre, A. M., Gateau-Olesker, A., Vass, G., Gero, S. D., Guthrie, R. D., Voelter, W. and Breitmaier, E. (1972). *Tetrahedron Lett.*, 5163
38. Mussini, P., Orsini, F. and Pelizzoni, F. (1973). *Tetrahedron Lett.*, 4849
39. Gough, J. L., Guthrie, J. P. and Stothers, J. B. (1972). *Chem. Commun.*, 979
40. Olah, G. A., Liang, G., Mateescu, G. D. and Riemenschneider, J. L. (1973). *J. Amer. Chem. Soc.*, **95**, 8698
41. Masamune, S., Sakai, M., Ona, H. and Jones, A. J. (1972). *J. Amer. Chem. Soc.*, **94**, 8956
42. Hart, H. and Kuzuya, M. (1973). *Tetrahedron Lett.*, 4123
43. Hogeveen, H. and Kwart, P. W. (1973). *Tetrahedron Lett.*, 1665
44. Goldstein, M. J. and Kline, S. A. (1973). *J. Amer. Chem. Soc.*, **95**, 935
45. Bates, R. B., Brenner, S., Cole, C. M., Davidson, E. W., Forsythe, G. D., McCombs, D. A. and Roth, A. S. (1973). *J. Amer. Chem. Soc.*, **95**, 926
46. Polonsky, J., Baskevitch, Z., Cagnoli-Bellavita, N., Ceccherelli, P., Buckwalter, B. L. and Wenkert, E. (1972). *J. Amer. Chem. Soc.*, **94**, 4369

47. Kluender, H., Bradley, C. H., Sih, C. J., Fawcett, P. and Abraham, E. P. (1973). *J. Amer. Chem. Soc.*, **95**, 6149
48. Neuss, N., Nash, C. H., Baldwin, J. E., Lemke, P. A. and Grutzner, J. B. (1973). *J. Amer. Chem. Soc.*, **95**, 3797
49. Milavetz, B., Kakinuma, K., Rinehart, K. L., jr., Rolls, J. P. and Haak, W. J. (1973). *J. Amer. Chem. Soc.*, **95**, 5793
50. Battersby, A. R., Hunt, E. and McDonald, E. (1973). *J. Chem. Soc. Chem. Commun.*, 442
51. Scott, A. I., Townsend, C. A., Okada, K., Kajiwara, M., Whitman, P. J. and Cushley, R. J. (1972). *J. Amer. Chem. Soc.*, **94**, 8267
52. Scott, A. I., Townsend, C. A. and Cushley, R. J. (1973). *J. Amer. Chem. Soc.*, **95**, 5759
53. Battersby, A. R., Ihara, M., McDonald, E., Stephenson, J. R. and Golding, B. T. (1973). *J. Chem. Soc. Chem. Commun.*, 404
54. Tanabe, M., Hamasaki, T., Suzuki, Y. and Johnson, L. F. (1973). *J. Chem. Soc. Chem. Commun.*, 212
55. Seto, H., Sato, T. and Yonehara, H. (1973). *J. Amer. Chem. Soc.*, **95**, 8461
56. Seto, H., Cary, L. W. and Tanabe, M. (1973). *J. Chem. Soc. Chem. Commun.*, 867
57. Tanabe, M., Suzuki, K. T. and Janowski, W. C. (1973). *Tetrahedron Lett.*, 4723
58. Crow, W. D. and Paddon-Row, M. N. (1972). *J. Amer. Chem. Soc.*, **94**, 4746
59. Whitesides, G. M. and Lewis, D. W. (1971). *J. Amer. Chem. Soc.*, **93**, 5914
60. Goering, H. L., Eikenberg, J. N. and Koerner, G. S. (1971). *J. Amer. Chem. Soc.*, **93**, 4913
61. Fraser, R. R., Petit, M. A., and Miskow, M. (1972). *J. Amer. Chem. Soc.*, **94**, 3254
62. Kainosho, M., Ajisaka, K., Pirkle, W. H. and Beare, S. D. (1972). *J. Amer. Chem. Soc.*, **94**, 5924
63. Bleaney, B. (1972). *J. Magn. Resonance*, **8**, 91
64. Bleaney, B., Dobson, C. M., Levine, B. A., Martin, R. B., Williams, R. J. P. and Xavier, A. V. (1972). *J. Chem. Soc. Chem. Commun.*, 791
65. Tori, K., Yoshimura, Y., Kainosho, M. and Ajisaka, K. (1973). *Tetrahedron Lett.*, 3127
66. Uebel, J. J., Pacheco, C. and Wing, R. M. (1973). *Tetrahedron Lett.*, 4383
67. Cramer, R. E. and Seff, K. (1972). *J. Chem. Soc. Chem. Commun.*, 400
68. Uebel, J. J. and Wing, R. M. (1972). *J. Amer. Chem. Soc.*, **94**, 8910
69. Wing, R. M., Uebel, J. J. and Anderson, K. K. (1973). *J. Amer. Chem. Soc.*, **95**, 6046
70. Briggs, J. M., Moss, G. P., Randall, E. W. and Sales, K. D. (1972). *J. Chem. Soc. Chem. Commun.*, 1180
71. Cramer, R. E. and Dubois, R. (1973). *J. Chem. Soc. Chem. Commun.*, 936
72. Hawkes, G. E., Liebfritz, D., Roberts, D. W. and Roberts, J. D. (1973). *J. Amer. Chem. Soc.*, **95**, 1659
73. Johnson, B. F. G., Lewis, J., McArdle, P. and Norton, J. R. (1972). *J. Chem. Soc. Chem. Commun.*, 535
74. Reuben, J. and Leigh, J. S., jr. (1972). *J. Amer. Chem. Soc.*, **94**, 2789
75. Bhacca, N. S., Selbin, J. and Wander, J. D. (1972). *J. Amer. Chem. Soc.*, **94**, 8719
76. Tori, K., Yoshimura, Y., Kainosho, M. and Ajisaka, K. (1973). *Tetrahedron Lett.*, 1573
77. Cushley, R. J., Anderson, D. R. and Lipsky, S. R. (1972). *J. Chem. Soc. Chem. Commun.*, 636
78. Hirayama, M., Edagawa, E. and Hanyu, Y. (1972). *J. Chem. Soc. Chem. Commun.*, 1343
79. Chalmers, A. A. and Pachler, K. G. R. (1972). *Tetrahedron Lett.*, 4033
80. Gansow, O. A., Loeffler, P. A., Davis, R. E., Willcott, M. R. and Lenkinski, R. E. (1973). *J. Amer. Chem. Soc.*, **95**, 3389, 3390
81. Wolkowski, Z. W., Beauté, C. and Jantzen, R. (1972). *J. Chem. Soc., Chem. Commun.*, 619
82. Shapiro, B. L., Johnston, M. D., Godwin, A. D., Proulx, T. W. and Shapiro, M. J. (1972). *Tetrahedron Lett.*, 3233
83. LaMar, G. N. and Faller, J. W. (1973). *J. Amer. Chem. Soc.*, **95**, 3817
84. Nautsch, D. F. S. (1973). *J. Amer. Chem. Soc.*, **95**, 1688
85. Faller, J. W. and LaMar, G. N. (1973). *Tetrahedron Lett.*, 1381

86. Gibb, V. G., Armitage, I. M., Hall, L. D. and Marshall, A. G. (1972). *J. Amer. Chem. Soc.*, **94,** 8919
87. Sanders, J. K. M., Hanson, S. W. and Williams, D. H. (1972). *J. Amer. Chem. Soc.*, **94,** 5325
88. ApSimon, J. W., Beierbeck, H. and Fruchier, A. (1973). *J. Amer. Chem. Soc.*, **95,** 939
89. Shapiro, B. L. and Johnston, M. D., jr. (1972). *J. Amer. Chem. Soc.*, **94,** 8185
90. Ghotra, J. S., Hart, F. A., Moss, G. P. and Staniforth, M. L. (1973). *J. Chem. Soc. Chem. Commun.*, 113
91. Reuben, J. (1973). *J. Amer. Chem. Soc.*, **95,** 3534
92. Evans, D. F. and Wyatt, M. (1973). *J. Chem. Soc. Chem. Commun.*, 339
93. Grotens, A. M., Backus, J. J. M., Pijpers, F. W. and de Boer, E. (1973). *Tetrahedron Lett.*, 1467
94. Liu, K.-T. (1972). *Tetrahedron Lett.*, 5039
95. Shapiro, B. L., Johnston, M. D., jr. and Towns, R. L. R. (1972). *J. Amer. Chem. Soc.*, **94,** 4381
96. Patrick, T. B. and Patrick, P. H. (1972). *J. Amer. Chem. Soc.*, **94,** 6230
97. Bentrude, W. G., Tan, H.-W. and Yee, K. C. (1972). *J. Amer. Chem. Soc.*, **94,** 3264
98. Sato, T. and Goto, K. (1973). *J. Chem. Soc. Chem. Commun.*, 494
99. Armitage, I. M., Hall, L. D., Marshall, A. G. and Werbelow, L. G. (1973). *J. Amer. Chem. Soc.*, **95,** 1437
100. Demarco, P. V., Cerimele, B. J., Crane, B. W. and Thakkar, A. L. (1972). *Tetrahedron Lett.*, 3539
101. Willcott, M. R., Lenkinski, R. E. and Davis, R. E. (1972). *J. Amer. Chem. Soc.*, **94,** 1742
102. Davis, R. E. and Willcott, M. R. (1972). *J. Amer. Chem. Soc.*, **94,** 1744
103. ApSimon, J. W. and Beierbeck, H. (1973). *Tetrahedron Lett.*, 581
104. Wing, R. M., Early, T. A. and Uebel, J. J. (1972). *Tetrahedron Lett.*, 4153
105. Davis, R. E., Willcott, M. R., Lenkinski, R. E., Doering, W. von E., and Birladenau, L. (1973). *J. Amer. Chem. Soc.*, **95,** 6846
106. Willcott, M. R., Davis, R. E., Faulkner, D. J. and Stallard, M. O. (1973). *Tetrahedron Lett.*, 3967
107. Hudson, C. E. and Bauld, N. L. (1973). *J. Amer. Chem. Soc.*, **95,** 3822
108. Angerman, N. S., Danyluk, S. S. and Victor, T. A. (1972). *J. Amer. Chem. Soc.*, **94,** 7137
109. Montaudo, G., Librando, V., Caccamese, S. and Maravigna, P. (1973). *J. Amer. Chem. Soc.*, **95,** 6365
110. Servis, K. L. and Barker, D. J. (1973). *J. Amer. Chem. Soc.*, **95,** 3392
111. Crombie, L., Findley, D. A. R. and Whiting, D. A. (1972). *Tetrahedron Lett.*, 4027
112. Connolly, J. D., Harding, A. E. and Thornton, I. M. S. (1973). *J. Chem. Soc. Chem. Commun.*, 1320
113. Paddon-Row, M. N., Watson, P. L. and Warrener, R. N. (1973). *Tetrahedron Lett.*, 1033
114. Berson, J. A. and Holder, R. W. (1973). *J. Amer. Chem. Soc.*, **95,** 2037
115. Wilson, S. R. and Turner, R. B. (1973). *J. Chem. Soc. Chem. Commun.*, 557
116. Tangerman, A. and Zwanenburg, B. (1973). *Tetrahedron Lett.*, 79
117. Graves, R. E. and Rose, P. I. (1973). *J. Chem. Soc. Chem. Commun.*, 630
118. Maskasky, J. E. and Kenney, M. E. (1973). *J. Amer. Chem. Soc.*, **95,** 1443
119. Maskasky, J. E., Mooney, J. R. and Kenney, M. E. (1972). *J. Amer. Chem. Soc.*, **94,** 2132
120. Gouedard, M., Gaudemer, F. and Gaudemer, A. (1973). *Tetrahedron Lett.*, 2257
121. Schiemenz, G. P. and Hansen, H. P. (1973). *Angew. Chem. Int. Ed. Engl.*, **12,** 400
122. Schurig, V. (1972). *Tetrahedron Lett.*, 3297
123. Denning, R. G., Rossotti, F. J. C. and Sellars, P. J. (1973). *J. Chem. Soc. Chem. Commun.*, 381
124. Dongala, E. B., Solladié-Cavallo, A. and Solladié, G. (1972). *Tetrahedron Lett.*, 4233
125. Pasto, D. J. and Borchardt, J. K. (1973). *Tetrahedron Lett.*, 2517
126. Reich, C. J., Sullivan, G. R. and Mosher, H. S. (1973). *Tetrahedron Lett.*, 1505
127. Gerlach H. and Zagalak, B. (1973). *J. Chem. Soc. Chem. Commun.*, 274

128. Hoyer, G. A., Rosenberg, D., Rufer, C. and Seeger, A. (1972). *Tetrahedron Lett.*, 985
129. Dale, J. A. and Mosher, H. S. (1973). *J. Amer. Chem. Soc.*, **95**, 512
130. Baxter, C. A. R. and Richards, H. C. (1972). *Tetrahedron Lett.*, 3357
131. Abraham, R. J. and Siverns, T. M. (1973). *Org. Magn. Resonance*, **5**, 253
132. Kainosho, M., Ajisaka, K., Pirkle, W. H. and Beare, S. D. (1972). *J. Amer. Chem. Soc.*, **94**, 5924
133. Fraser, R. R., Petit, M. A. and Miskow, M. (1972). *J. Amer. Chem. Soc.*, **94**, 3253
134. McKinney, J. D., Matthews, H. B. and Wilson, N. K. (1973). *Tetrahedron Lett.*, 1896
135. Acheson, R. M. and Selby, I. A. (1973). *J. Chem. Soc. Chem. Commun.*, 537
136. Altona, C. and Sundaralingam, M. (1973). *J. Amer. Chem. Soc.*, **95**, 2333
137. Cistaro, C., Merlini, L., Mondelli, R. and Nasini, G. (1972). *J. Chem. Soc. Chem. Commun.*, 785
138. Barfield, M. and Gearhart, H. L. (1973). *J. Amer. Chem. Soc.*, **95**, 641
139. Solkien, V. N. and Bystrov, V. F. (1973). *Tetrahedron Lett.*, 2261
140. Pease, L. G., Deber, C. M. and Blout, E. R. (1973). *J. Amer. Chem. Soc.*, **95**, 258
141. Mark, V. (1974). *Tetrahedron Lett.*, 299
142. Barfield, M. and Sternhell, S. (1972). *J. Amer. Chem. Soc.*, **94**, 1905
143. Davis, D. B. and Abu Khaled, M. (1973). *Tetrahedron Lett.*, 2829
144. Gerig, J. T. and Macleod, R. S. (1973). *J. Amer. Chem. Soc.*, **95**, 5725
145. Agranat, I., Rabinovitz, M., Gosnay, I. and Weitzen-Dagan, A. (1972). *J. Amer. Chem. Soc.*, **94**, 2889
146. Wetzel, R. B. and Kenyon, G. L. (1972). *J. Amer. Chem. Soc.*, **94**, 9230
147. Benezra, C. (1973). *J. Amer. Chem. Soc.*, **95**, 6890
148. Bentrude, W. G. and Tan, H.-W. (1973). *J. Amer. Chem. Soc.*, **95**, 4666
149. Bushweller, C. H. and Brunelle, J. A. (1973). *J. Amer. Chem. Soc.*, **95**, 5949
150. Bock, K., Lundt, I. and Pedersen, C. (1973). *Tetrahedron Lett.*, 1037
151. Jennings, W. B., Boyd, D. R., Watson, C. G., Becker, E. D., Bradley, R. B. and Jerina, D. M. (1972). *J. Amer. Chem. Soc.*, **94**, 8501
152. Jerome, F. R. and Servis, K. L. (1972). *J. Amer. Chem. Soc.*, **94**, 5896
153. Doddrell, D., Jordan, D., Riggs, N. V. and Wells, P. R. (1972). *J. Chem. Soc. Chem. Commun.*, 1158
154. Haemers, M., Ottinger, R., Zimmermann, D. and Reisse, J. (1973). *Tetrahedron Lett.*, 2241
155. Gray, G. A. and Cremer, S. E. (1972). *J. Chem. Soc. Chem. Commun.*, 367
156. Breen, J. J., Featherman, S. I., Quin, L. D. and Stocks, R. C. (1972). *J. Chem. Soc. Chem. Commun.*, 657
157. Featherman, S. I. and Quin, L. D. (1973). *Tetrahedron Lett.*, 1955
158. Simonnin, M. P., Lequan, R.-M. and Wehrli, F. W. (1972). *J. Chem. Soc. Chem. Commun.*, 1204
159. Wetzel, R. B. and Kenyon, G. L. (1973). *J. Chem. Soc. Chem. Commun.*, 287
160. Mantsch, H. H. and Smith, I. C. P. (1972). *Biochem. Biophys. Res. Commun.*, **46**, 808
161. Lapper, R. D., Mantsch, H. H. and Smith, I. C. P. (1972). *J. Amer. Chem. Soc.*, **94**, 6243
162. Weigert, F. J. and Roberts, J. D. (1972). *J. Amer. Chem. Soc.*, **94**, 6021
163. Summerhays, K. D. and Maciel, G. E. (1972). *J. Amer. Chem. Soc.*, **94**, 8348
164. Freeman, R. and Hill, H. D. W. (1970). *J. Chem. Phys.*, **53**, 4103
165. Hall, L. D. and Preston, C. (1972). *J. Chem. Soc. Chem. Commun.*, 1319
166. Chachaty, C., Wolkowski, Z., Piriou, F. and Lukacs, G. (1973). *J. Chem. Soc. Chem. Commun.*, 951
167. Doddrell, D. and Allerhand, A. (1971). *J. Amer. Chem. Soc.*, **93**, 1558
168. Levy, G. C., Cargioli, J. D. and Anet, F. A. L. (1973). *J. Amer. Chem. Soc.*, **95**, 1527
169. Allerhand, A. and Doddrell, D. (1971). *J. Amer. Chem. Soc.*, **93**, 2777
170. Allerhand, A., Doddrell, D. and Komorosky, R. (1971). *J. Chem. Phys.*, **55**, 189
171. Allerhand, A., Doddrell, D., Glushko, V., Cochran, D. W., Wenkert, E., Lawson, P. J. and Gurd, F. R. N. (1971). *J. Amer. Chem. Soc.*, **93**, 544
172. Wehrli, F. W. (1973). *J. Chem. Soc. Chem. Commun.*, 379
173. Nakanishi, K., Gullo, V. P., Miura, I., Govindachari, T. R. and Viswanathan, N. (1973). *J. Amer. Chem. Soc.*, **95**, 6473
174. Homer, J., Dudley, A. R. and McWhinnie, W. R. (1973). *J. Chem. Soc. Chem. Commun.*, 893

175. Noggle, J. H. and Schirmer, R. E. (1971). *The Nuclear Overhauser Effect: Chemical Applications* (New York: Academic Press)
176. Schirmer, R. E., Davis, J. P., Noggle, J. H. and Hart, P. A. (1972). *J. Amer. Chem. Soc.*, **94**, 2561
177. Son, T.-D., Guschlbauer, W. and Guéron, M. (1972). *J. Amer. Chem. Soc.*, **94**, 7903
178. Schirmer, R. E. and Noggle, J. H. (1972). *J. Amer. Chem. Soc.*, **94**, 2947
179. Smith, R. V. and Stocklinski, A. W. (1973). *Tetrahedron Lett.*, 1819
180. Garbesi, A., Barbarella, G. and Fava, A. (1973). *J. Chem. Soc. Chem Commun.*, 155
181. Martin, R. H., Eyndels, Ch. and Defay, N. (1972). *Tetrahedron Lett.*, 2731
182. Horibe, I., Tori, K., Takeda, K. and Ogino, T. (1973). *Tetrahedron Lett.*, 735
183. Tori, K., Horibe, I., Tamura, Y. and Tada, H. (1973). *J. Chem. Soc. Chem. Commun.*, 620
184. Brown, R. T., Heatley, F., Moorcroft, D. and Ladd, J. A. (1973). *J. Chem. Soc. Chem. Commun.*, 459
185. Cheng, H. N. and Gutowsky, H. S. (1972). *J. Amer. Chem. Soc.*, **94**, 5505
186. Tanny, S. R., Pickering, M. and Springer, C. S., jr. (1973). *J. Amer. Chem. Soc.*, **95**, 6227
187. Kessler, H. and Molter, M. (1973). *Angew. Chem. Int. Ed. Engl.*, **12**, 1011
188. Mannschreck, A., Jonas, V. and Kolb, B. (1973). *Angew. Chem. Int. Ed. Engl.*, **12**, 582
189. Mannschreck, A., Jonas, V. and Kolb, B. (1973). *Angew Chem. Int. Ed. Engl.*, **12**, 909
190. Yang, C. S. C. and Liu, R. S. H. (1973). *Tetrahedron Lett.*, 4811
191. Booth, H. (1973). *J. Chem. Soc. Chem. Commun.*, 945
192. Anet, F. A. L., Chmurny, G. N. and Krane, J. (1973). *J. Amer. Chem. Soc.*, **95**, 4423
193. Bernard, M., Sauriol, F. and St. Jacques, M. (1972). *J. Amer. Chem. Soc.*, **94**, 8624
194. Lambert, J. B., Mixan, C. E. and Johnson, D. H. (1973). *J. Amer. Chem. Soc.*, **95**, 4634
195. Booth, H. and Griffiths, D. V. (1973). *J. Chem. Soc. Chem. Commun.*, 666
196. de Pessemier, F., Tavernier, D. and Anteunis, M. (1973). *J. Chem. Soc. Chem. Commun.*, 594
197. Coll, J. C., Crist, De L. R., Barrio, M. del C. G. and Leonard, N. J. (1972). *J. Amer. Chem. Soc.*, **94**, 7092
198. Kiefer, E. F., Levek, T. J. and Bopp, T. T. (1972). *J. Amer. Chem. Soc.*, **94**, 4751
199. Glazer, E. S., Knorr, R., Ganter, C. and Roberts, J. D. (1972). *J. Amer. Chem. Soc.*, **94**, 6026
200. Anet, F. A. L. and Basus, V. J. (1973). *J. Amer. Chem. Soc.*, **95**, 4424
201. Anet, F. A. L. and Kozerski, L. (1973). *J. Amer. Chem. Soc.*, **95**, 3407
202. Dale, J., Ekeland, T. and Krane, J. (1972). *J. Amer. Chem. Soc.*, **94**, 1389
203. Anet, F. A. L. and Degen, P. J. (1972). *J. Amer. Chem. Soc.*, **94**, 1390
204. Anet, F. A. L. and Degen, P. J. (1972). *Tetrahedron Lett.*, 3613
205. Crossley, R., Downing, A. P., Nógrádi, M., Braga de Oliviera, A., Ollis, W. D. and Sutherland, I. O. (1973). *J. Chem. Soc. Perkin Trans. I*, 205
206. Montecalvo, D., St Jacques, M. and Wasylishen, R. (1973). *J. Amer. Chem. Soc.*, **95**, 2023
207. Buchanan, G. W. (1972). *Tetrahedron Lett.*, 665
208. Senkler, G. H., jr., Gust, D., Riccobono, P. X. and Mislow, K. (1972). *J. Amer. Chem. Soc.*, **94**, 8626
209. Anet, F. A. L., Cheng, A. K. and Wagner, J. J. (1972). *J. Amer. Chem. Soc.*, **94**, 9250
210. Noe, E. A. and Roberts, J. D. (1972). *J. Amer. Chem. Soc.*, **94**, 2020
211. Noe, E. A., Wheland, R. C., Glazer, E. S. and Roberts, J. D. (1972). *J. Amer. Chem. Soc.*, **94**, 3488
212. Anet, F. A. L., Cheng, A. K. and Krane, J. (1973). *J. Amer. Chem. Soc.*, **95**, 7877
213. Brickwood, D. J., Ollis, W. D. and Stoddart, J. F. (1973). *J. Chem. Soc. Chem. Commun.*, 638
214. Fujita, S., Hirano, S. and Nozaki, H. (1972). *Tetrahedron Lett.*, 403
215. Rosenfeld, S. and Keehn, P. M. (1973). *Tetrahedron Lett.*, 4021
216. Sherrod, S. A. and Boekelheide, V. (1972). *J. Amer. Chem. Soc.*, **94**, 5513
217. Sherrod, S. A. and daCosta, R. L. (1973). *Tetrahedron Lett.*, 2083
218. Nakamura, M., Oki, M. and Nakanishi, H. (1973). *J. Amer. Chem. Soc.*, **95**, 7169
219. Anderson, J. E. and Rawson, D. I. (1973). *J. Chem. Soc. Chem. Commun.*, 830
220. Anderson, J. E., Franck, R. W. and Mandella, W. L. (1972). *J. Amer. Chem. Soc.*, **94**, 4608

4
X-ray Crystallography

A. F. CAMERON
University of Glasgow

4.1 INTRODUCTION   99

4.2 STUDIES OF INTRAMOLECULAR BONDING BY X-RAY CRYSTALLOGRAPHY 100

4.3 MOLECULAR COMPLEXES AND INTERMOLECULAR BONDING
     INTERACTIONS   106

4.4 MOLECULAR CONFORMATIONS   110

4.5 BIOLOGICAL STUDIES   116

4.1 INTRODUCTION

In the period covered by the present article, the trend of studying non-heavy-atom derivatives of organic molecules by direct methods of structure solution using accurate data collected by automatic diffractometer techniques has developed to the point of being almost commonplace. One result is that detailed molecular dimensions may now be discussed with a confidence which reflects the greater accuracy of such analyses. Moreover, where they are available, modern computers can now handle crystallographic calculations with such speed that for many laboratories the limiting factor in producing crystallographic results is often the number of sets of data which can be collected in a given period of time. One consequence of these developments is that the volume of results which has appeared in the present period of literature coverage is such that no more than a representative sample can be chosen for inclusion in this article. It has therefore been decided to choose analyses, the results of which exemplify the importance of crystallographic results in the areas of intramolecular bonding, intermolecular associations, molecular conformations and biological activity. Analyses undertaken simply to resolve ambiguous or unknown molecular structures have not been included unless the results have proved to be of additional interest in one of the above contexts. This approach parallels that adopted by the author in Organic Chemistry Series One, Volume 1.

4.2 STUDIES OF INTRAMOLECULAR BONDING BY X-RAY CRYSTALLOGRAPHY

As mentioned above, the widespread use of direct methods of structure solution combined with diffractometer techniques of data collection have together resulted in accurate determinations of the dimensions of complex organic molecules which would previously have been studied as heavy-atom derivatives. One consequence is that variations in $C(sp^3)$—$C(sp^3)$ bond lengths are now being recognised as significant. Such variations are thought to result both from repulsion between non-bonded atoms, and also from orbital hybridisation effects which increase the carbon single-bond radius with decreasing s-character in the bond[1]. Thus in the 'symmetrical cedrone' (1)[2]

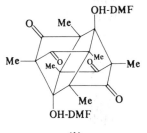

(1)

which is a dimeric oxidation product of trimethylphloroglucinol, the bonds C(1)—C(2) [1.597(3) Å], C(1)—C(6) [1.597(3) Å] and C(1)—C(4') [1.577(3) Å] are all significantly 'lengthened', and the extensions may be associated with eclipsing of substituents bonded to the involved atoms. The photodimer of 5,6,7,8-tetrahydro-2-quinolone (2)[3] provides another example in which the length of the intermonomer bonds is 1.623(3) Å. In this case it is suggested

(2)   (3)

that the facile reversal of the dimer to monomeric units is reflected in the 'long' bonds which effect the dimerisation. Similarly, in the heterocyclic cage compound (3)[4], the $C(sp^3)$—$C(sp^3)$ bond forming the junction of the fused piperazine rings has a length of 1.574(3) Å. These three cases represent only a few of the many examples in which relatively long C—C single bonds result from steric effects.

Even more remarkable, however, are the results obtained for two substituted 1,6-methano[10]annulenes (4a)[5] and (4b)[6]. Both chemical and n.m.r.

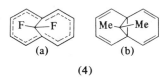

(a)   (b)

(4)

evidence had indicated that whereas the 11,11-difluoro derivative (4a) should be an annulene, the 11,11-dimethyl analogue (4b) should tend to possess a bisnorcaradiene structure. The crystallographic studies of both compounds support these postulates, the difluoro compound clearly being an annulene with a non-planar ring and some alternation of bond lengths, while the dimethyl compound tends towards the bisnorcaradiene structure. However, the tendency of (4b) towards the bisnorcaradiene formulation is in itself of considerable interest, since although the geometry is clearly non-aromatic, with the potentially aromatic bonds having lengths in the range 1.34–1.46 Å, the C(1)—C(6) 'bond' exhibits lengths of 1.82(1) and 1.77(1) Å in two crystallographically independent molecules. It is suggested that this bond is so weak that it is susceptible to intermolecular forces. Another example, spiro(indene-1,7'-norcaradiene) (5)[7], represents a fluctuating norcaradiene–cycloheptatriene system in solution. However, in the solid state the molecule is clearly a norcaradiene in which the C(10)—C(15) bond has a length of 1.520(5) Å.

(5)   (6)

8,8-Dichlorotricyclo[3.2.1.0^{1,5}]octane (6)[8] represents yet another case in which there are unusual geometrical features associated with a bridgehead–bridgehead bond. In this molecule, the length of the C(1)—C(5) bond is 1.572(15) Å, and both bridgehead atoms display inverted geometry with all four interatomic vectors lying on the same side of a plane. In detail, each of the bridgehead atoms is displaced almost 0.1 Å *outwards* from the plane described by each set of three nearly trigonal non-bridgehead neighbours. However, the observed geometry of this molecule probably highlights the dangers implicit in assuming that interatomic vectors are always coincident with bonds, since the angles between corresponding bond hybrids are found to lie in the range 101–116°, thus leading to a model more closely akin to that expected for four-coordinate carbon than might be inferred simply from connecting atomic centres.

In Series One of this review, mention was made of the considerable number of thiathiophthens and analogous compounds which were being studied to investigate S—S and S—O bonds of lengths intermediate between single-bond values and non-bonded interactions. In the intervening period there has appeared a description[9] of CNDO/2 calculations for 6a-thiathiophthens, in which the effects of methyl and phenyl substitution on the S—S bonding have been examined relative to the unsubstituted molecule (7) [S—S 2.350 Å].

(7)

The energy curves predict that whereas 2-methyl substitution causes lengthening of the S—S bond in the same ring, 3-methyl substitution results in shortening of the same bond. 2-Phenyl substitution also causes lengthening of the associated S—S bond, which varies, however, with the angle of twist of the phenyl ring, while 3-phenyl substitution has little effect. The theoretical results are generally in agreement with experimentally observed values, including those of the recently described examples (8a)[10] [S(1)—S(2) 2.425, S(2)—S(3) 2.301 Å], (8b)[11] [2.375, 2.266 Å], (8c)[12] [2.255, 2.398 Å], (8d)[13] [2.350, 2.348 Å], (8e)[14] (two-fold symmetry) [both bonds 2.303 Å], (9a)[15]

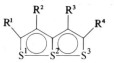

(8) (a) R¹ = Me, R² = R³ = R⁴ = H
   (b) R¹ = SH, R² = NH₂, R³ = H, R⁴ = Ph
   (c) R¹ = R⁴ = Ph, R² = Me, R³ = H
   (d) R¹ = R³ = H, R² = Ph, R⁴ = C₆H₄NMe₂-p
   (e) R¹ = R⁴ = Ph, R² = R³ = Me

(9) (a) n = 3, R¹ = R² = Ph
   (b) n = 2, R¹ = R² = Ph

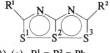

(10) (a) R¹ = R² = Ph
    (b) R¹ = R² = NHPh

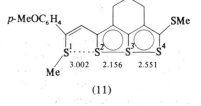

(11)

[2.329, 2.288 Å] and (9b)[16] (mirror symmetry) [both bonds 2.351 Å]. Diaza analogues of the thiathiophthens such as (10a)[17] [S(1)—S(2) 2.319, S(2)—S(3) 2.328 Å] and (10b)[18] [S(1)—S(2) 2.225, S(2)—S(3) 2.475 Å] also tend to have dimensions in agreement with those which would be predicted on a theoretical basis. In contrast to the predictable behaviour of the thiathiophthens, compounds in which there are linear sequences of four sulphur atoms are apparently able to exhibit a variety of bonding patterns. Thus in (11)[19] and (12)[20] the sequences S(2)—S(3)—S(4) constitute well-defined thiathiophthen systems, with only weak interactions involving S(1) in each case. However, in (13)[21], S(1)—S(2) and S(3)—S(4) form recognisable dithiole systems which

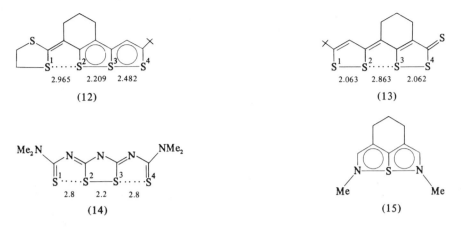

(12)   (13)

(14)   (15)

interact only weakly with each other, while in (14)[22] there is a central 1,2,4-dithiazole ring which interacts weakly with the two flanking sulphur atoms.

Bonding situations analogous to those of the thiathiophthens may also occur in examples where the sulphur atoms forming the linear sequence are replaced by nitrogen, oxygen or selenium. Thus in (15)[23], N—S bonds of 1.901(5) and 1.948(5) Å are observed, while in (16)[24] and (17)[25] the S—O bonds lie in the range 2.284–2.577 Å. Selenium-containing examples include

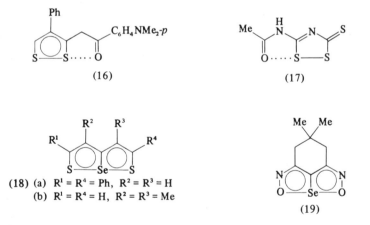

(16)   (17)

(18) (a) $R^1 = R^4 = Ph$, $R^2 = R^3 = H$
    (b) $R^1 = R^4 = H$, $R^2 = R^3 = Me$

(19)

(18a)[26] [Se—S 2.419 and 2.433 Å], (18b)[27] [Se—S 2.414 Å] and (19)[28] [Se—O average 2.023 Å]. Long S—O and Se—O interactions are not, however, limited to the thiathiophthens and their analogues. For example, examinations of o-carboxyphenyl methyl sulphoxide (20)[29] and o-carboxyphenyl-methylselenium oxide (21)[29] reveal that whereas the sulphur compound is as shown with the hydrogen atom located on the carboxyl group, and an S···O interaction of 2.777 Å, the selenium analogue is cyclised with transfer of the hydrogen atom, although the Se—O bond [2.378 Å] is extremely long. Moreover, approximately linear arrangements of the three-atom O—S—O sequence are not limited to analogues of thiathiophthens. Thus the O—S—O

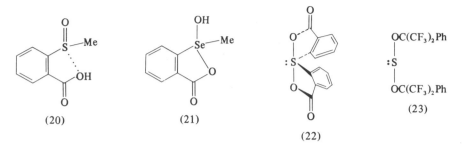

(20)   (21)   (22)   (23)

sequences of (22)[30] and (23)[31] are described by O—S—O valency angles of 178.5° and 175.1°, respectively, while the S—O bonds [1.83–1.92 Å] are extremely long. Whereas the angle between the phenyl rings in (22) is 106.9°, the geometry of the sulphur atom in (23) is described as a trigonal bipyramid. It is suggested that the O—S—O bonding in (23) is effected by two-electron three-centre molecular orbitals.

Phosphorus-containing compounds may also adopt a variety of geometries, exemplified by (24)[32], in which the phosphorus atom is six-coordinate and possesses octahedral geometry. In the case of (25)[33] the geometry of the phosphorus atom is intermediate between trigonal-bipyramidal and square-pyramidal.

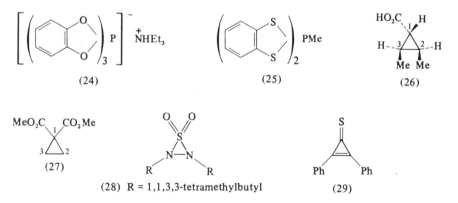

(24)   (25)   (26)

(27)   (28) R = 1,1,3,3-tetramethylbutyl   (29)

Studies of cyclopropane derivatives have revealed some features of interest. In both (26)[34] and (27)[35], the C(2)—C(3) bonds [typically 1.473(5) Å] are considerably shorter than either the C(1)—C(2) or C(1)—C(3) bonds [typically 1.541(6) Å], and it is suggested that this effect is associated with a polar effect resulting from the orientation of the carbonyl oxygen atoms with respect to the ring. Heterocyclic analogues of cyclopropane which have been examined include (28)[36] in which the N—N bond [1.67 Å] is extremely long, and (29)[37] in which the shortness of the C(1)—C(2) bond indicates that the molecule is not well-defined in terms of a delocalised zwitterionic structure.

Partially delocalised five-membered heterocyclic compounds are represented by the concerted study of the triazole derivatives (30a–d)[38–41] and (31)[42]. This study was undertaken to investigate the effects of a variety of substituents on the ring geometry, and the results are presented in Table 4.1. From

Table 4.1 Dimensions* of various triazole derivatives

Bond	Compound				
	(30a)[38]	(30b)[39]	(30c)[40]	(30d)[41]	(31)[42]
N(1)—N(2)	1.405(5)	1.390(3)	1.412(4)	1.370(2)	1.372(3)
N(2)—C(3)	1.298(5)	1.317(3)	1.301(4)	1.287(2)	1.316(3)
C(3)—N(4)	1.311(5)	1.345(3)	1.364(4)	1.367(2)	1.373(3)
N(4)—C(5)	1.311(5)	1.368(3)	1.391(4)	1.381(2)	1.373(3)
C(5)—N(1)	1.298(5)	1.333(3)	1.315(4)	1.340(2)	1.358(3)

* In Å

these it may be concluded that interchange of the 5-NH$_2$ and 5-SH substituents has a marked effect only on the N(1)—C(5) and N(4)—C(5) bonds. Moreover, the salt (30c) differs from the other sulphur derivatives only in the length of the C(5)—S$^-$ bond [1.724(4) Å], which is slightly longer than the C(5)—SH bonds [average 1.671(3) Å] of the other molecules. Considering only those derivatives which have been examined, the geometry of the 1,2,4-triazole ring is apparently relatively insensitive to interchange of substituents.

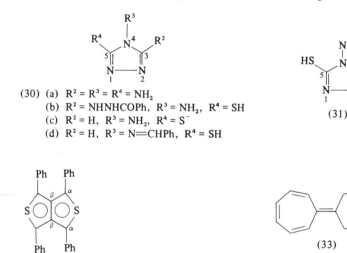

(30) (a) R^2 = R^3 = R^4 = NH$_2$
      (b) R^2 = NHNHCOPh, R^3 = NH$_2$, R^4 = SH
      (c) R^2 = H, R^3 = NH$_2$, R^4 = S$^-$
      (d) R^2 = H, R^3 = N=CHPh, R^4 = SH

(31)

(32)

(33)

Aromaticity is also a feature which has been extensively studied by x-ray analysis. Thus, tetraphenylthieno[3,4-c]thiophene (32)[43] represents a 10π-electron aromatic system, possessing a centrosymmetric molecule with a planar central moiety. The lengths of the C—S bonds are 1.706 Å, while the C$^\alpha$—C$^\beta$ and C$^\beta$—C$^\beta$ bonds have lengths of 1.407 and 1.452 Å, respectively. The non-alternant hydrocarbon heptafulvalene (33)[44] has also been examined to provide details of the molecular geometry. The molecules are centrosymmetric and possess a non-planar conformation which may be described as S-shaped, with a largest deviation of 0.35 Å from the best molecular plane.

There is almost complete alternation of bond lengths between near-single [1.464 Å] and near-double [1.338 Å] values, while the linking bond has a length of 1.379 Å. There is also strong correlation between bond length and bond torsion, with the strain of the seven-membered rings being relieved almost exclusively in the longer bonds. Tropone (34a)[45], 3-azidotropone

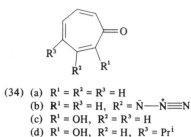

(34) (a) $R^1 = R^2 = R^3 = H$
(b) $R^1 = R^3 = H$, $R^2 = \bar{N}\!-\!\overset{+}{N}\!\equiv\!N$
(c) $R^1 = OH$, $R^2 = R^3 = H$
(d) $R^1 = OH$, $R^2 = H$, $R^3 = Pr^i$

(34b)[46], tropolone (34c)[47] and 4-isopropyltropolone (34d)[48] are all, however, planar. They are also said to show [with the exception of bonds involving C(1)] less marked alternation of bond lengths than does heptafulvalene. Typical 'double bond' and 'single bond' lengths are 1.364 and 1.413 Å, respectively, while the carbonyl bond typically has a length of 1.261 Å.

4.3   MOLECULAR COMPLEXES AND INTERMOLECULAR BONDING INTERACTIONS

A considerable number of analyses have been devoted to investigations of the interactions which may exist between molecules in the solid state. The analyses of 3,6-diphenyl-*sym*-tetrazine[49] and *p*-diethynylbenzene[50] are of relevance to such studies, since in each case the solid-state packing arrangements were predicted by calculation of theoretical crystal structures using potential energy curves for the interactions between neighbouring molecules.

The crystal structures of molecular complexes involving aromatic molecules have also attracted considerable interest, since the complex formation often involves a degree of charge-transfer interaction. Examples are provided by the 1:1 complexes anthracene : 1,2,4,5-tetracyanobenzene[51] and 2,4,6-trinitrobenzene : 3-formylbenzothiophene[52]. In both complexes the constituent molecules are stacked alternatively in infinite columns, the interplanar spacings in the former being *ca.* 3.42 Å, and in the latter 2.97–3.22 Å. 7,7,8,8,-Tetracyanoquinodimethane (TCNQ) also forms a large variety of complexes and salts, the solid-state structures of which invariably involve a degree of charge-transfer interaction. Examples of 1:1 complexes involving TCNQ include the associations with benzidine[53], dibenzo-*p*-dioxin[54], phenazine[55], *N*-dimethyldihydrophenazine[56] and 1,10-phenanthroline[57]. The range [3.22–3.46 Å] of interplanar spacings in these complexes indicates a variety of charge transfer interactions, usually within infinite columns of alternately stacked components. The 1,10-phenanthroline complex is unusual in that the overlapping donor–acceptor pairs exist as separate entities rather than as mixed stacks. The salts of TCNQ exhibit packing arrangements and charge-transfer interactions which are distinctly different from those of the complexes. Examples which have been examined include [N,N'-dibenzyl-4,4'-bipyridi-lium)$^{2+}$ (TCNQ)$_4^{2-}$]^{58}, [(1,1'-ethylene-2,2'-bipyridilium)$^{2+}$ (TCNQ)$_2^{2-}$]^{59}, [(N-(n-propyl)quinolinium)$^+$ (TCNQ)$_2^-$]^{60}, [(morpholinium)$^+$ (TCNQ)$^-$]^{61} and [(morpholinium)$_2^{2+}$ (TCNQ)$_3^{2-}$]^{62}. The structures of these salts consist of infinite columns of TCNQ entities, the columns being separated by a sub-structure of anions. Charge-transfer interactions are constrained to take place between small numbers of TCNQ entities within the TCNQ columns. The subunits within the TCNQ columns may be diads or centrosymmetric tetrads, with typical interplanar spacings between the four TCNQ entities of a tetrad being 3.16 and 3.24 Å. The inter-tetrad spacing [typically 3.62 Å] usually indicates only minimal charge-transfer interaction between tetrads.

Charge-transfer interactions are not, however, constrained to take place by face-to-face stacking of aromatic or delocalised molecules and ionic species. Thus a feature of the crystal structure of (35)^{63} is the existence of short O$^-$· · ·Br [2.796 Å] intermolecular contacts, which are interpreted as

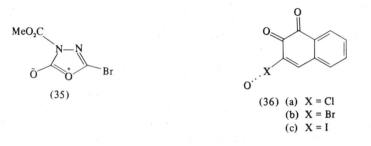

(35)

(36) (a) X = Cl
     (b) X = Br
     (c) X = I

indicating charge-transfer interaction. Such oxygen· · ·halogen interactions have been further investigated by i.r. and x-ray studies of (36a–c)^{64}, all of which display linear C—X· · ·O arrangements in which the strength of the bond increases with acceptor strength from chlorine through bromine to iodine. Such interactions are also possible between nitrogen and halogen atoms, examples being provided by the crystal structures of (37a–c)^{65} [Cl$^-$· · · N 3.36 Å, Br$^-$· · ·N 3.68 Å, I$^-$· · ·N 3.84 Å] and (38)^{66} [C≡N· · ·Cl ca. 3.00 Å]. Interactions involving cyano groups are also evident in the structures of

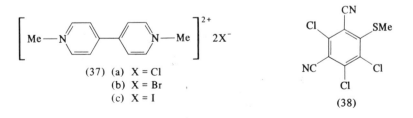

(37) (a) X = Cl
     (b) X = Br
     (c) X = I

(38)

2,4,6-trichloro-^{67} [C≡N· · ·Cl 3.22 Å] and 2,4,6-tribromo-benzonitrile^{67} [C≡N· · ·Br 3.06 Å], while the crystal structure of p-iodotoluene^{68} contains short I· · ·I [4.06 Å] contacts. The crystal structures of certain acetylenic compounds may also be of interest. In the case of di-iodoacetylene^{69} the molecules are arranged such that each molecule points at both ends towards the middles of two adjacent molecules to give short I··· ‖$\genfrac{}{}{0pt}{}{C}{C}$ [C···I *ca.* 3.4 Å] contacts, which are interpreted as donor-(C≡C)···acceptor-I interactions. Similarly, the ethynyl hydrogen of but-3-ynoic acid[70] points towards the mid-point of an adjacent triple bond to give a short C—H··· ‖$\genfrac{}{}{0pt}{}{C}{C}$ contact.

X-Ray crystallographic methods are also used to investigate solid-state reactions, molecular rearrangements and other similar transformations. In such studies, different degrees of transformation must be recognised. Thus phase changes may involve a simple reorientation and slight rearrangement of molecular position, with perhaps an accompanying adjustment of molecular conformation, while solid-state molecular rearrangements will involve both the molecule → product chemical reaction and also the formation of a solid-state structure for the products. In addition, separate molecules may dimerise in the solid state, the dimerisation being accompanied by the formation of a crystal structure for the dimers.

Studies of phase changes are represented by the analyses of both the yellow and white forms of 3,6-dichloro-2,5-dihydroxyterephthalate[71], for which a yellow → white transformation is observed. Both crystalline modifications crystallise in the triclinic space group PĪ, and although the molecules of the yellow phase are planar, centrosymmetric and are characterised by an internal hydrogen bond, in the white phase the molecules prove to be non-planar and to be linked by intermolecular hydrogen bonding. Comparison of the two crystal structures reveals that the yellow → white transformation requires a change from intra- to inter-molecular hydrogen bonding, a change in conformation from planar to non-planar, and also a flipping of every aromatic ring through 180°. Yet another example is provided by cyclohexane[72], which forms two crystalline phases at low temperatures, the higher-temperature Phase I [stable 186–220 K] undergoing an isothermal transition to the lower-temperature Phase II [stable <186 K] at 186 K. Whereas Phase I is cubic, Phase II is monoclinic, although an examination of the molecular arrangement within the cubic cell of Phase I reveals that a monoclinic cell, very similar to that of Phase II, may be *constructed.* Moreover, to produce the

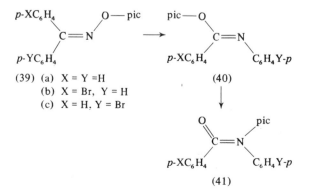

(39) (a) X = Y = H
    (b) X = Br, Y = H
    (c) X = H, Y = Br

(40)

(41)

Phase II molecular arrangement within the *constructed* monoclinic cell requires only one major molecular reorientation.

Molecular rearrangements which have been studied in the solid state include the Beckmann rearrangement of the benzophenone oxime O-picryl ethers (39a–c) to form [via the intermediates (40a–c)] the respective picryl anilides (41a–c). The crystallographic investigation[73] (Table 4.2) included full

Table 4.2 Crystallographic data for the derivatives (39a–c) and (41a–c)

Compound	Space group	Volume per molecule/Å³	Packing fraction*
(39a)	C2/c or Cc	474(4)	0.95
(41a)	P2₁/c	455(3)	
(39b)	P2₁/c	498(3)	0.96
(41b)	P2₁/c	476(3)	
(39c)	P2₁/c	492(3)	0.99
(41c)	Pnma or Pna2₁	484(4)	

* The packing fraction is defined as $1 - \Delta/\tau$, where Δ is the difference in molecular volumes, and τ is the volume of the larger molecule

analyses of (39b) and (39c), and reveals that whereas the conversions (39a) → (41a) and (39b) → (41b) give rise to a contraction of only *ca.* 5% in cell volume, the conversion (39c) → (41c) produces no measurable contraction. In those rearrangements where the molecular shapes and interplanar spacings of the starting materials suggest minimal disruption of the crystal, the products separate as microcrystallites which have no net orientation with respect to the parent crystal. However, it is possible to construct models of the products which have space-filling characteristics similar to those of the reactants, and this may well account for the tendency of the products to remain in the matrix of the starting material. In yet another example, phenyl-azotribenzoylmethane (42a) undergoes a solid-state reaction and rearrangement to form α-phenylazo-β-benzoyloxybenzalacetophenone (43a) and diphenyl triketone *sym*-benzoylphenylhydrazone (44a). In conjunction with a

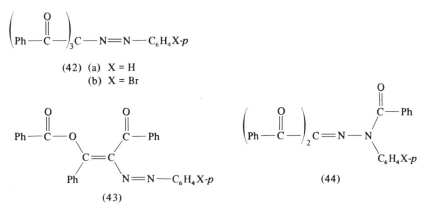

$$\left(\underset{\substack{\| \\ O}}{Ph-C} \right)_3 C-N{=}N-C_6H_4X\text{-}p$$

(42) (a) X = H
     (b) X = Br

(43)

(44)

study of the kinetics of the reaction, analyses of (42a)[74], (43b)[74], (44a)[75] and (44b)[75] have reported. In the case of (42a) there are no *intermolecular* distances between migrating centres of less than 4.79 Å, while for (43b) the stereochemistry about the N=N bond is *trans* and the migrating carbonyl group [to form (44b)] and the receiving *p*-bromophenylazo group are also *trans* with respect to the C=C bond.

The photodimerisation of (45) has also been studied, although it proved possible only to complete a full analysis of the dimer (46)[76]. However, both

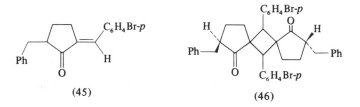

(45)   (46)

monomer and dimer crystallise in the same space group *Pbca* with unit cells of similar volumes (<3.5% difference). It would seem that the small changes in cell dimensions are sufficient to allow the dimerisation to proceed without drastic reorganisation of the molecular arrangement. In another example the analysis of (47)[77], which undergoes a solid-state photodimerisation, has been

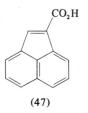

(47)

reported. The packing of the molecules of (47) is such that the distance between neighbouring molecules is favourable for a topochemical reaction.

Several studies of hydrogen bonding have revealed the presence of hydrated oxonium ions. Thus, the crystal structure of picrylsulphonic acid tetrahydrate[78] contains the $H_5O_2^+$ [$H_3O^+ \cdot H_2O$] ion, which is completed by an O—H$\cdots$O hydrogen bond [O$\cdots$O 2.429 Å]. Both normal[79] and deuterated[79] sulphuric acid also contain the $H_5O_2^+$ [O$\cdots$O 2.431 Å] and analogous $D_5O_2^+$ [O$\cdots$O 2.426 Å] ions, respectively. The $H_7O_3^+$ [$H_3O^+ \cdot 2H_2O$] ion is found in the crystal structures of 2,5-dibromobenzenesulphonic acid trihydrate[80], 2,5-dichlorobenzenesulphonic acid trihydrate[81] and perchloric acid trihydrate[82]. In each case the hydrogen bonds completing the $H_7O_3^+$ ions are similar in length to those completing the $H_5O_2^+$ ions.

4.4 MOLECULAR CONFORMATIONS

The techniques of x-ray crystallography are widely used in the study of molecular conformations, although in this context care must be exercised when extrapolating from results which pertain strictly to the solid state. The conformations of aromatic derivatives have been studied in some depth, usually to investigate distortions which may result from the introduction of bulky substituents. Thus, although 1,2,4,5-tetra-butylbenzene (48)[83] is

(48)

centrosymmetric and, with the exception of the methyl groups, planar, the substitution pattern results in considerable distortions in the valency angles C(1)C(6)C(5), C(4)C(5)C(10) and C(5)C(4)C(9), which are all increased to *ca.* 130°, while the angles C(3)C(4)C(5) and C(4)C(5)C(6) are both decreased to *ca.* 115°. In another examples, 2,3,5,6-tetrachloro-(4-methylthio)benzonitrile[84], it is reported that the effect of six substituents on a benzene ring is to induce slight, but significant, non-planarity, the deviations of the substituents from the mean molecular plane lying in the range −0.10 to +0.16 Å. A heterocyclic derivative, hexamethylmelamine (49)[85], forms complexes with

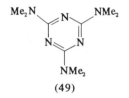

(49)

other aromatic molecules in which the molecule is planar. The present analysis demonstrates that the molecule retains its planarity in the uncomplexed state.

Hexahelicene and several of its derivatives have also been examined. In the case of hexahelicene (50)[86] itself, the conformation has been described simply

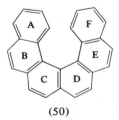

(50)

in terms of the interplanar angles between the individual least-squares planes of the rings A, B, C, D, E and F [A:B 9.8°, B:C 15.2°, C:D 14.4°, D:E 15.2°, E:F 11.5°, A:F 58.5°]. However, since each of the rings is significantly non-planar, perhaps a more meaningful description of the conformation is that given for 2-methylhexahelicene (51)[87], in which the planar portions 1, 2, 2', 3 and 3' are identified [1:2 14.8°, 1:2' −13.6°, 2:2' −25.2°, 2:3 −43.1°, 3:3'

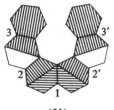

(51)

−53.5°]. In this case the angle between the least-squares planes of the terminal rings A and F is 54.8°. In this study the authors also point out that the range of peripheral bond lengths [1.322–1.446 Å] is quite different from the range of core bond lengths [1.405–1.466 Å]. The shortening of the outer bonds is most marked for those bonds which are parallel to and immediately opposite core bonds. The analysis of (−)-2-bromohexahelicene[88], and its subsequent conversion to (−)-hexahelicene, together prove that (−)-hexahelicene possesses the left-handed helical absolute configuration.

Biphenyls which have been studied include p-nitrobiphenyl[89] and 3'-iodobiphenyl-4-carboxylic acid[90]. In the former case the dihedral angle between the planes of the phenyl rings is 33°, while a value of 30.4° is reported for the iodo derivative. A bridged example, 5H,8H-dibenzo[df][1,2]dithiocin (52)[91], exhibits a dihedral angle of 57.3° between the phenyl planes, while the dihedral angle at the disulphide bond is 54.3°. Although 2,7-di-t-butylpyrene (53)[92], which is in some respects analogous to the bridged biphenyls, proves to have a planar and centrosymmetric aromatic nucleus, the tetrahydro derivative (54)[93] presents a rather more complex conformational problem.

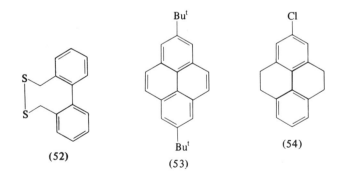

(52)   (53)   (54)

In this case the crystal structure contains two crystallographically independent molecules, one of which is disordered. The disordered molecule possesses a conformation in which the phenyl–phenyl dihedral angle is close to zero, while the conformation of the ordered molecule is such that the phenyl–phenyl dihedral angle is 16.9°. These conformational differences result from the different possible relative orientations of the ethylene bridges. It is also pointed out that in addition to a rotation about the biphenyl linking bond, the dihedral angle between the phenyl rings also results to a significant extent from a 'bowing' of the molecule. This latter effect is possibly somewhat analogous to the distortions observed in the cyclophanes, of which the quadruple-layered molecule (55)[94] is an extreme example. The inner benzene

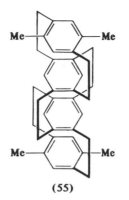

(55)

rings of this molecule are distorted into twisted-boat conformations in which the individual atoms of these rings are displaced by as much as 0.16 Å from planarity.

Cyclobutane rings are apparently able to adopt a range of puckered conformations. Thus in (56)[95] and (57)[96] the puckerings of the rings are described by dihedral angles of 156° and 152.6°, respectively, between the two distinct three-atom planes of each ring. However, the cyclobutane ring of (58)[97] is considerably less puckered, with a dihedral angle of 172.3°. The molecule of

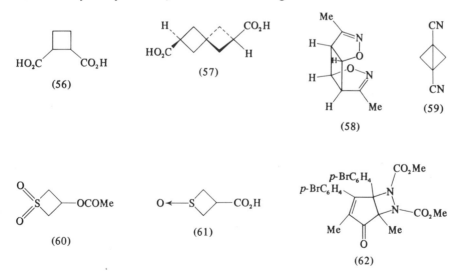

(59)[98] is a remarkable example of a bridged cyclobutane ring, in which the dihedral angle between the cyclopropane portions is 126.4°, and the length of the bridging bond is 1.502(6) Å. Heterocyclic analogues of cyclobutane which have been examined include (60)[99] [dihedral angle 162°], (61)[100] [dihedral angle 153°] and (62)[101] [dihedral angle 173°].

There have been several studies of t-butyl derivatives of cyclohexane, including *cis*-4-t-butylcyclohexane-1-carboxylic acid[102], *cis*-[103] and *trans*-4-t-butylcyclohexyl toluene-*p*-sulphonate[104], and *trans*-3-t-butylcyclohexyl toluene-*p*-sulphonate[105]. In all of these examples the cyclohexane rings adopt chair conformations in which the t-butyl groups occupy equatorial positions. The chair conformations are, however, considerably flattened in comparison with less heavily substituted derivatives of cyclohexane. The flattening is reflected both in the values of the torsion angles for the ring bonds [typical average |54.8°|, typical range |51.4°|–|57.2°|], and also in the values of the endocyclic valency angles, which typically average 111.5°. Rather surprisingly, *cis*-2-chloro-4-t-butylcyclohexanone[106], in which the mean torsion angle is |56.5°|, the torsion angle range is |51.2°|–|58.0°| and the mean internal valency angle is 111.3°, exhibits a conformation similar to the conformations of the previous t-butyl derivatives. Heterocyclic analogues of cyclohexane include (63)[107], in which the six-membered ring adopts a chair conformation with the fluorine atoms occupying axial positions. All six S–N bonds have experimentally identical lengths [1.593 Å], this feature being taken to indicate electron delocalisation. The S—N—S and N—S—N valency angles are 123.2° and 112.6°, respectively. Cyclotrimethylenetrinitramine (64)[108] and 2,4,6-trimethyl-1,3,5-triselenane (65)[109] also adopt their conformations. In the case of (64) the internal valency angles at the nitrogen atoms average 114.5°,

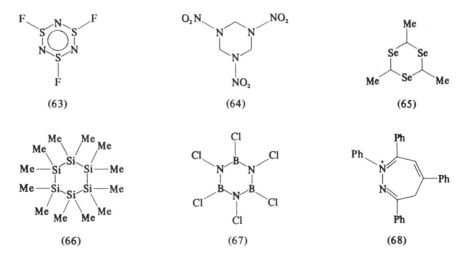

(63)   (64)   (65)

(66)   (67)   (68)

while for (65) the average internal valency angle at the selenium atoms is 101.1°. Dodecamethylcyclohexasilane (66)[110] exhibits an almost perfect chair conformation in which the average Si—Si bond length is 2.338(4) Å and the average Si—Si—Si valency angle is 111.9(4)°. The mean torsion angle for the ring bond is |53.5°|. In contrast to the chair conformations of the previous heterocyclic derivatives, hexachloroborazine (67)[111] possesses a planar ring, although the chlorine atoms are displaced by ±0.04 Å from the plane of the ring. Moreover, the B—N bond lengths alternate from 1.398 to 1.451 Å.

Compounds with larger rings which have been examined include 3,5,7-triphenyl-4*H*-1,2-diazepine picrate (68)[112], in which the 1,2-diazepine ring adopts a boat conformation with the methylene group at the prow. In contrast, the seven-membered ring of *syn*-8,8-dichloro-4-phenyl-3,5-dioxa-bicyclo[5.1.0]octane[113] adopts the chair conformation (69). The eight-membered ring of (70)[114] adopts a twist-boat conformation in which there is a dihedral angle of *ca.* 74° between the mean molecular planes containing the olefinic double bonds. In this case, solution n.m.r. studies indicate that the same conformation also predominates in solution. The 14-membered ring of (71)[115] adopts a conformation very similar to that of cyclo-tetradecane, which may be derived from the diamond structure. However, although the 2/*m* symmetry of cyclotetradecane is lost, the inversion centre is maintained, with the carbon and oxygen atoms of the independent unit defining the two almost perpendicular planes [83°] formed by C(1), O(1), C(2), C(3), C(4), and C(1), O(2), C(4), C(5).

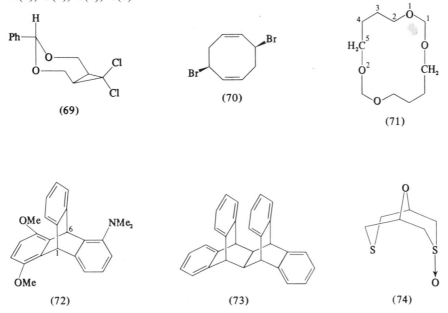

(69)

(70)

(71)

(72)

(73)

(74)

The two bicyclo[2.2.2]octyl derivatives (72)[116] and (73)[117] show the same general indications of molecular strain, although in each case the phenyl rings retain their planarity. Thus, at the bridgehead atoms C(1) and C(6) in (72), and at the corresponding atoms in (73), the endocyclic valency angles are *ca.* 104°. Moreover, consideration of the valency angles at each bridge reveals that those angles exocyclic with respect both to the central bicyclic system, and also with respect to the phenyl rings, are increased to *ca.* 126°, while the endocyclic valency angles at each bridge have values *ca.* 114°. A heterocyclic analogue of bicyclo[3.3.1]nonane (74)[118], adopts a chair–chair conformation such that the intramolecular distance between the sulphur atoms is only 3.17 Å. The C—S—C valency angles average 97.0°, and the C—S—O valency angles average 103.9°. The length of the S—O bond is 1.508 Å.

4.5 BIOLOGICAL STUDIES

The crystallographic method is widely applied to studies of molecules which are of biological significance. Such molecules may be natural products, or compounds possessing biological activity of clinical importance. Although a discussion of crystallographic studies of proteins and of other natural macromolecules is beyond the scope of this chapter, nevertheless, a considerable number of analyses have been devoted to studies of the individual components of such large molecules, and it is pertinent to provide a brief mention of this work. Simple peptides which have been examined include glycyl-L-alanine hydrochloride[119], which possesses a non-planar peptide linkage. The conformation of this molecule (see Ref. 120) is such that the torsion angle ω has a value of 10.2°, while ψ and ϕ have values of 11 and 108°, respectively. In another study[121], N-(bromoacetyl)-L-phenylalanyl-L-phenylalanine ethyl ester and its chloroacetyl analogue have been shown to be isostructural. The conformation of the backbone of this dipeptide, which is a substrate of pepsin, is described by the angles ϕ_1, ϕ_2, ψ_1, ψ_{T1} and ψ_{T2} which have values of -102, -122, $+106$, -133 and $+53°$, respectively.

Bases which have been studied include 5,6-dihydro-1-methyl-4-thiouracil (75)[122], which adopts a half-chair conformation with the molecules in the ketothione form. The molecular packing of (75) is such that the two crystallographically independent molecules form N—H$\cdots$O [2.840, 2.829 Å] hydrogen-bonded pairs, the molecules within each pair being twisted ca. 33° with respect to each other. The photopolymers produced by u.v. irradiation of proteins and bases have also excited considerable interest. Thus,

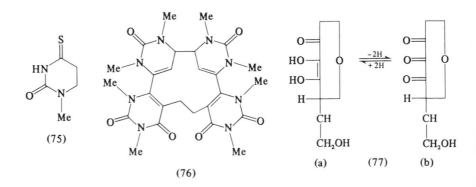

6,4'-(pyrimidin-2'-one)thymine is a photoproduct isolated in minute quantities from u.v.-irradiated DNA. Moreover, irradiation of an aqueous solution of this adduct results in yet another adduct, the structure of which has been shown to be (76)[123]. Since (76) contains four pyrimidine nuclei, it may be regarded as a pyrimidine phototetramer.

Vitamin C is a compound which is of topical interest and controversy. It is, however, generally accepted that an important property of vitamin C is its function as a reducing agent, the redox process having been described by the reversible reaction (77a) → (77b). However, the formulation (77b)

for dehydro-L-ascorbic acid has been criticised on the grounds that three adjacent carbonyl groups are incompatible with the colourless nature of the material. Since suitable crystals of dehydro-L-ascorbic acid are difficult to obtain, an analysis of dimerised dehydro-L-ascorbic acid has in the meantime been carried out. The dimer proves to have the molecular structure (78)[124],

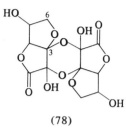

(78)

and while the structure of the monomer is still in doubt, there is nevertheless some further support for the contention that (77) is not a true representation of the redox properties of L-ascorbic acid. The molecule (78) possesses two-fold symmetry, and the γ-lactone rings are planar with the exception of C(3), which is 0.19 Å distant from the plane of the other atoms. The furanose ring has an irregular envelope conformation in which C(6) is 0.55 Å removed from the plane of the other four atoms, and the dioxan ring has a (two-fold) symmetrical twist-boat conformation.

In the steroid field there have been several analyses of 5,6-unsaturated androstenes to study the effect of the double bond on the molecular conformations. The compounds which have been examined are: 3β-p-bromobenzoyloxyandrost-5-en-17-one (79)[125]; 3β-chloroandrost-5-en-17β-ol (80a)[126]; 3β-acetoxy-17α-iodoandrost-5-ene (80b)[127]; and the fluorinated pyran derivative (81)[128]. In each case ring A adopts a chair conformation [typical torsion-angle range |46°|–|59°|], ring B adopts a half-chair conformation

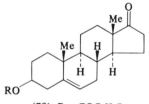

(79) R = COC$_6$H$_4$Br-p

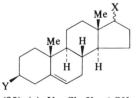

(80) (a) Y = Cl, X = β-OH
(b) Y = CO$_2$Me, X = α-I

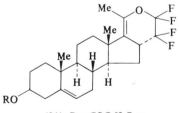

(81) R = COC$_6$H$_4$Br-p

[typically |0°|–|60°|] and ring C adopts a chair form [typically |42°|–|61°|]. Although rings A, B and C of the four compounds are very similar, there are variations in the ring D conformations. Thus in (79), ring D adopts a half-chair form [|1°|–|38°|], while in the other three molecules, β-envelope conformations of varying distortions are observed.

In addition to the types of analyses reported above, considerable effort is now being devoted to relating detailed molecular geometry and geometrical changes to biological activities. Thus 11-*cis*-retinal (82)[129] acts as a photochemical sensor in visual systems, and in the dark is covalently linked to

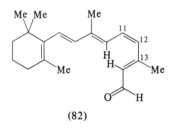

(82)

proteins (opsins) in the retina. The primary event in the visual excitation process is the conversion of the 11-*cis*-isomer into all-*trans*-retinal. In an effort to provide information which will contribute towards a detailed understanding of the visual process, analyses of both isomers have been undertaken. The chain of all-*trans*-retinal[130] proves to be extended, although it is markedly curved both within its general plane, and is also slightly bent normal to the plane. The main feature of the side-chain of 11-*cis*-retinal is the significantly non-zero torsion angle about the C(12)—C(13) single bond, such that the conversion of the 11-*cis* into the all-*trans* isomer involves both a rotation of 180° about the C(11)—C(12) double bond, and also a rotation of *ca.* 141° about the C(12)—C(13) bond.

Yet another example is provided by the analyses of diphenylhydantoin (83)[131] and diazepam (84)[132], which although chemically different, show

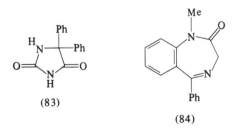

(83)

(84)

similar anticonvulsant drug activities. Examinations of the conformations of both molecules reveal that despite the chemical differences, the space-filling characteristics are very similar. In particular, the relative orientations of the phenyl rings and the carbonyl groups in the two molecules are directly comparable. This observation leads to a tentative and preliminary suggestion that this feature may well be pertinent to anticonvulsant activity.

In Series One of this report, mention was made of the importance attached to the conformations of several molecules containing ethanolamine side chains, in relation to their biological activities. These molecules included the bronchodilators isoproterenol sulphate[133], (−)-ephedrine hydrochloride[134], adrenaline hydrochloride[135] and amphetamine[136], all of which were characterised both by extended conformations of the side chains and also by *cisoid* relationships of the ammonium and hydroxyl groups which thus come into close proximity. These studies have now been considerably extended by analyses of several more ethanolamine-containing biologically active molecules, including the adrenergic stimulants Th 1179 (85)[137] and its diastereoisomer Th 1165a[138], Alupent (86)[139] and Salbutamol (87)[140]. Whereas Th

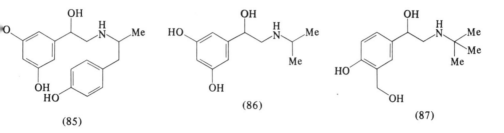

(85)   (86)   (87)

1165a is a potent stimulant of adrenergic receptors, and exhibits a conformation in which the β-ethanolamine side chain is fully extended, and the amine and hydroxyl functions are *cis*, the diastereoisomer Th 1179 is 9–20 times less active. The analysis of Th 1179 reveals that the configurational change results in a folded conformation for the side chain. As a result of these two studies in particular, it is tentatively suggested that, with reference to the process of muscle relaxation, it would appear that stimulation of a particular receptor site is directly related to (a) the molecular cross-section and thickness of the molecular chain, and (b) to the linearity or extended nature of the molecular chain. The analyses of Alupent and Salbutamol reveal conformations which lend support to this hypothesis. Additional analyses of (−)-adrenaline hydrogen (+)-tartrate[141], and of (−)-ephedrine dihydrogen phosphate[142], reveal that (−)-adrenaline has the absolute configuration (88), and that the conformation of (−)-ephedrine (89) in the

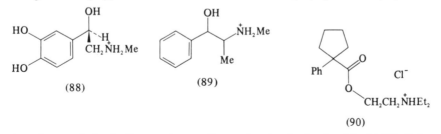

(88)   (89)   (90)

phosphate salt is similar to the conformation in the hydrochloride[134]. Other ethanolamine- and ethylamine-containing biologically active molecules which have been examined include 2-diethylaminoethyl-1-phenylcyclopentanecarboxylate hydrochloride (90)[143], mescaline hydrochloride (91)[144] and the antihistamine (92)[145].

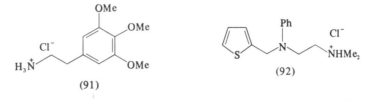

(91)

(92)

Whereas *p*-methoxybenzenesulphon-*p*-anisidide (93a)[146] and the *N*-isopropyl derivative (93b)[147] both display oestrogenic activity, the former molecule is the most active of the series, while the latter is the least active. In (93a) the MeO·· ·OMe distance of 8.72 Å is less than the distance of 11 Å regarded as necessary for oestrogenic activity, although it is pointed out that this distance could be achieved by rotation about the S—N bond. Moreover,

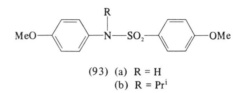

(93) (a) R = H
(b) R = Pr^i

in the two crystallographically independent molecules of (93b), the MeO·· · OMe distances are 9.45 and 9.17 Å, and although the conformation differs from that of (93a) by a rotation of 25° about the S—N bond, the presence of the isopropyl group will hinder the further rotation which would be necessary to achieve the O·· ·O 11 Å requirement.

The various configurational isomers of promedol alcohol, which are potent analgesics, have been subject to further examination. The rhombohedral form of (±)-β-promedol alcohol (94)[148] contains molecules whose conformation is substantially identical to that of the monoclinic form. The piperidine ring adopts a skew-chair form [N(1), C(2), C(4), C(5) planar; C(3) and

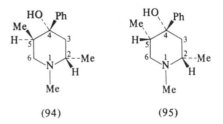

(94)

(95)

C(6) +0.53 and −0.66 Å distant from the plane], in which the phenyl group is axial and the three methyl groups are equatorial. (±)α-Promedol alcohol (95)[149] also proves to have a slightly distorted chair conformation [C(2), C(3), C(5), C(6) approximately planar; N(1) 0.68 Å and C(4) −0.60 Å distant from the plane] in which the phenyl group is more nearly perpendicular (axial) than in the least-active γ-isomer. As a result of the several analyses of the promedol alcohols, and the known potencies which are γ < α < β, various conclusions can be drawn regarding the structural requirements for activity.

Firstly, the highest potency is achieved with the *cis*-(5-methyl/4-phenyl) configuration. Secondly, where the configuration is *trans*-(5-methyl/4-phenyl), the potency is greatest for the *trans*-(2-methyl/4-phenyl) isomer, and least for the *cis*-(2-methyl/4-phenyl) isomer. Moreover, potency does not seem to be influenced by the orientation of the equatorial phenyl ring relative to the piperidine ring.

Insecticides are currently exciting considerable interest, and in response there has been reported a series of analyses of various compounds which possess insecticidal activities. The compounds which have been studied include *p,p'*-DDT (96a)[150], *o,p'*-DDT (96b)[150], and the DDT analogues (97)[151],

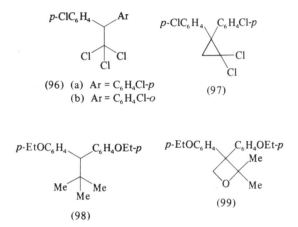

(96) (a) Ar = C_6H_4Cl-*p*
      (b) Ar = C_6H_4Cl-*o*

(97)

(98)

(99)

(98)[151], (99)[152], endrin[153] and aldrin[153]. The structural interest in these compounds stems from the theoretical space-filling requirements of fitting organic molecules into the interspaces of nerve membrane, and the relationship of these requirements to insecticidal activity. Thus, for *p,p'*-DDT itself, the molecule is described in terms of the van der Waals' dimensions of the various plane projections, the diameter of the apex being 6.55 Å, the projected area of the trichloromethane group 2.96 Å^2, and the lengths of the two longer projections being 8.70 and 14.00 Å. The distance between the two electronegative chlorine atoms is 10.40 Å.

References

1. Alden, R. A., Krant, J. and Trayler, T. G. (1968). *J. Amer. Chem. Soc.*, **90**, 74
2. Beisler, J. A. and Silverton, J. V. (1972). *Acta Crystallogr.*, **B28**, 298
3. Brown, J. N., Towns, R. L. R. and Trefonas, L. M. (1971). *J. Amer. Chem. Soc.*, **93**, 7012
4. Gilardi, R. D. (1972). *Acta Crystallogr.*, **B28**, 742
5. Gramaccioli, C. M. and Simonetta, M. (1971). *Tetrahedron, Lett.*, 173
6. Bianchi, R., Mugnoli, A. and Simonetta, M. (1972). *J. Chem. Soc. Chem. Commun.*, 1073
7. Dreissig, W., Luger, P., Rewicki, D. and Tuchscherer, Ch. (1973). *Cryst. Struct. Commun.*, **2**, 197
8. Wiberg, K. B., Burgmaier, G. J., Shen, K.-W., La Placa, S. J., Hamilton, W. C. and Newton, M. D. (1972). *J. Amer. Chem. Soc.*, **94**, 7402

9. Hansen, L. K., Hordvik, A. and Saethre, L. J. (1972). *J. Chem. Soc. Chem. Commun.*, 222
10. Hordvik, A. and Saethre, L. J. (1972). *Acta Chem. Scand.*, **26**, 1729
11. Barnett, A. J., Beer, R. J. S., Karaoghlanian, B. V., Llaguno, E. C. and Paul, I. C. (1972). *J. Chem. Soc. Chem. Commun.*, 836
12. Hordvik, A., Sjølset, O. and Saethre, L. J. (1972). *Acta Chem. Scand.*, **26**, 1297
13. Hordvik, A. and Saethre, L. J. (1972). *Acta Chem. Scand.*, **26**, 3114
14. Hordvik, A., Sjølset, O. and Saethre, L. J. (1973). *Acta Chem. Scand.*, **27**, 379
15. Birknes, B., Hordvik, A. and Saerthe, L. J. (1972). *Acta Chem. Scand.*, **26**, 2140
16. Birknes, B., Hordvik, A. and Saethre, L. J. (1973). *Acta Chem. Scand.*, **27**, 382
17. Hordvik, A. and Milje, L. (1972). *J. Chem. Soc. Chem. Commun.*, 182
18. Hordvik, A. and Oftedal, P. (1972). *J. Chem. Soc. Chem. Commun.*, 543
19. Sletten, J. (1973). *Acta Chem. Scand.*, **27**, 229
20. Sletten, J. (1971). *Acta Chem. Scand.*, **25**, 3577
21. Sletten, J. (1972). *Acta Chem. Scand.*, **26**, 873
22. Oliver, J. E., Flippen, J. L. and Karle, J. (1972). *J. Chem. Soc. Chem. Commun.*, 1153
23. Hordvik, A. and Julshamn, K. (1972). *Acta Chem. Scand.*, **26**, 343
24. Hordvik, A. and Saethre, L. J. (1972). *Acta Chem. Scand.*, **26**, 899
25. Eide, G., Hordvik, A. and Saethre, L. J. (1972). *Acta Chem. Scand.*, **26**, 2140
26. Hordvik, A., Rimala, T. S. and Saethre, L. J. (1972). *Acta Chem. Scand.*, **26**, 2139
27. Hordvik, A., Rimala, T. S. and Saethre, L. J. (1973). *Acta Chem. Scand.*, **27**, 360
28. Llaguno, E. C. and Paul, I. C. (1972). *J. Chem. Soc. Perkin Trans. II*, 2001
29. Dahlen, B. (1973). *Acta Crystallogr.*, **B29**, 595
30. Kálmán, A., Sasvári, K. and Kapovits, I. (1973). *Acta Crystallogr.*, **B29**, 355
31. Paul, I. C., Martin, J. C. and Perozzi, E. F. (1971). *J. Amer. Chem. Soc.*, **93**, 6674; (1972). *J. Amer. Chem. Soc.*, **94**, 5010
32. Allcock, H. R. and Bissell, E. C. (1972). *J. Chem. Soc. Chem. Commun.*, 676
33. Eisenhut, M., Schmutzler, R. and Sheldrick, W. S. (1973). *J. Chem. Soc. Chem. Commun.*, 144
34. Luhan, P. A. and McPhail, A. T. (1972). *J. Chem. Soc. Perkin Trans. II*, 2372
35. de Jong, J. G. H. and Schenk, H. (1973). *Cryst. Struct. Commun.*, **2**, 25
36. Trefonas, L. M. and Cheung, L. D. (1973). *J. Amer. Chem. Soc.*, **95**, 636
37. Reed, L. L. and Schaefer, J. P. (1972). *J. Chem. Soc. Chem. Commun.*, 528
38. Seccombe, R. C. and Kennard, C. H. L. (1973). *J. Chem. Soc. Perkin Trans. II*, 1
39. Seccombe, R. C. and Kennard, C. H. L. (1973). *J. Chem. Soc. Perkin Trans. II*, 4
40. Seccombe, R. C., Tillack, J. V. and Kennard, C. H. L. (1973). *J. Chem. Soc. Perkin Trans. II*, 6
41. Seccombe, R. C. and Kennard, C. H. L. (1973). *J. Chem. Soc. Perkin Trans. II*, 9
42. Seccombe, R. C. and Kennard, C. H. L. (1973). *J. Chem. Soc., Perkin Trans. II*, 11
43. Glick, M. D. and Cook, R. E. (1972). *Acta Crystallogr.*, **B28**, 1336
44. Thomas, R. and Coppens, P. (1972). *Acta Crystallogr.*, **B28**, 1800
45. Barrow, M. J., Mills, O. S. and Filippini, G. (1973). *J. Chem. Soc. Chem. Commun.*, 66
46. Cruickshank, D. W. J., Filippini, G. and Mills, O. S. (1972). *J. Chem. Soc. Chem. Commun.*, 101
47. Shimanouchi, H. and Sasada, Y. (1973). *Acta Crystallogr.*, **B29**, 81
48. Derry, J. E. and Hamor, T. A. (1972). *J. Chem. Soc. Perkin Trans. II*, 694
49. Ahmed, N. A., Kitaigorodsky, A. I. and Sirota, M. I. (1972). *Acta Crystallogr.*, **B28** 2875
50. Ahmed, N. A. and Kitaigorodsky, A. I. (1972). *Acta Crystallogr.*, **B28**, 739
51. Tsuchiya, H., Marumo, F. and Saito, Y. (1972). *Acta Crystallogr.*, **B28**, 1935
52. Pascard, R. and Pascard-Billy, C. (1972). *Acta Crystallogr.*, **B28**, 1926
53. Ikemoto, I., Chikaishi, K., Yakushi, K. and Kuroda, H. (1972). *Acta Crystallogr.*, **B28**, 3502
54. Goldberg, I. and Shmueli, U. (1973). *Acta Crystallogr.*, **B29**, 432
55. Goldberg, I. and Shmueli, U. (1973). *Acta Crystallogr.*, **B29**, 440
56. Goldberg, I. and Shmueli, U. (1973). *Acta Crystallogr.*, **B29**, 421
57. Goldberg, I. and Shmueli, U. (1973). *Cryst. Struct. Commun.*, **2**, 175
58. Sundaresan, T. and Wallwork, S. C. (1972). *Acta Crystallogr.*, **B28**, 2474
59. Sundaresan, T. and Wallwork, S. C. (1972). *Acta Crystallogr.*, **B28**, 3065
60. Sundaresan, T. and Wallwork, S. C. (1972). *Acta Crystallogr.*, **B28**, 1163

61. Sundaresan, T. and Wallwork, S. C. (1972). *Acta Crystallogr.*, **B28**, 3507
62. Sundaresan, T. and Wallwork, S. C. (1972). *Acta Crystallogr.*, **B28**, 491
63. Seyferth, D., Shih, H.-M. and LaPrade, M. D. (1972). *J. Chem. Soc. Chem. Commun.*, 1036
64. Gaultier, J., Hauw, C. and Schvoerer, M. (1971). *Acta Crystallogr.*, **B27**, 2199
65. Russell, J. H. and Wallwork, S. C. (1972). *Acta Crystallogr.*, **B28**, 1527
66. Carter, D. R., Turley, J. W. and Boer, F. P. (1972). *Acta Crystallogr.*, **B28**, 3430
67. Carter, V. B. and Britton, D. (1972). *Acta Crystollogr.*, **B28**, 945
68. Ahn, C.-T., Soled, S. and Carpenter, G. B. (1972). *Acta Crystallogr.*, **B28**, 2152
69. Dunitz, J. D., Gehrer, H. and Britton, D. (1972). *Acta Crystallogr.*, **B28**, 1989
70. Benghiat, V. and Leiserowitz, L. (1972). *J. Chem. Soc. Perkin Trans. II*, 1772
71. Byrn, S. R., Curtin, D. Y. and Paul, I. C. (1972). *J. Amer. Chem. Soc.*, **94**, 890
72. Kahn, R., Fourme, R., André, D. and Renaud, M. (1973). *Acta Crystallogr.*, **B29**, 131
73. McCullough, J. D., Paul, I. C. and Curtin, D. Y. (1972). *J. Amer. Chem. Soc.*, **94**, 883
74. Pendergrass, D. B., Paul, I. C. and Curtin, D. Y. (1972). *J. Amer. Chem. Soc.*, **94**, 8722
75. Pendergrass, D. B., Curtin, D. Y. and Paul, I. C. (1972). *J. Amer. Chem. Soc.*, **94**, 8730
76. Whiting, D. A. (1971). *J. Chem. Soc. C*, 3397
77. Bouas-Laurent, H., Castellan, A., Desvergne, J. P., Dumartin, G., Courseille, C., Gaultier, J. and Hauw, C. (1972). *J. Chem. Soc. Chem. Commun.*, 1267
78. Lundgren, J.-O. (1972). *Acta Crystallogr.*, **B28**, 1684
79. Kjallman, T. and Olovsson, I. (1972). *Acta Crystallogr.*, **B28**, 1692
80. Lundgren, J.-O. (1972). *Acta Crystallogr.*, **B28**, 475
81. Lundgren, J.-O. and Lundin, P. (1972). *Acta Crystallogr.*, **B28**, 486
82. Almlöf, J. (1972). *Acta Crystallogr.*, **B28**, 481
83. Stam, C. H. (1972). *Acta Crystallogr.*, **B28**, 2715
84. Carter, D. R. and Boer, F. P. (1972). *J. Chem. Soc. Perkin Trans. II*, 2104
85. Bullen, G. J., Corney, D. J. and Stephens, F. S. (1972). *J. Chem. Soc. Perkin Trans. II*, 642
86. de Rango, C., Tsoucaris, G., Declerq, J. P., Germain, G. and Putzeys, J. P. (1973). *Cryst. Struct. Commun.*, **2**, 189
87. Frank, G. W., Hefelfinger, D. T. and Lightner, D. A. (1973). *Acta Crystallogr.*, **B29**, 223
88. Lightner, D. A. Hefelfinger, D. T., Powers, T. W., Frank, G. W. and Trueblood, K. N. (1972). *J. Amer. Chem. Soc.*, **94**, 3492
89. Casalone, G. Gavezzotti, A. and Simonetta, M. (1973). *J. Chem. Soc. Perkin Trans. II*, 342
90. Sutherland, H. H. and Mottram, M. J. (1972). *Acta Crystallogr.*, **B28**, 2212
91. Wahl, G. H., jun., Bordner, J., Harpp, D. N. and Gleason, J. G. (1972). *J. Chem. Soc. Chem. Commun.*, 985
92. Hazell, A. C. and Lomborg, J. G. (1972). *Acta Crystallogr.*, **B28**, 1059
93. Bear, C. A., Hall, D., Waters, J. M. and Waters, T. N. (1973). *J. Chem. Soc. Perkin Trans. II*, 314
94. Mizumo, H., Nishiguchi, K., Otsubo, T., Misumi, S. and Morimoto, N. (1972). *Tetrahedron, Lett.*, 4981
95. van der Helm, D., Hsu, I.-N. and Sime, J. M. (1972). *Acta Crystallogr.*, **B28**, 3109
96. Hulshof, L. A., Vos, A. and Wynberg, H. (1972). *J. Org. Chem.*, **37**, 1767
97. Andreetti, G. D., Bocelli, G. and Sgarabotto, P. (1973). *Cryst. Struct. Commun.*, **2**, 115
98. Johnson, P. L. and Schaefer, J. P. (1972). *J. Org. Chem.*, **37**, 2762
99. Andreetti, G. D., Bocelli, G. and Sgarabotto, P. (1972). *Cryst. Struct. Commun.*, **1**, 423
100. Abrahamsson, S. and Rehnberg, G. (1972). *Acta Chem. Scand.*, **26**, 494
101. Chieh, P. C., Mackay, D. and Wong, L. (1972). *J. Chem. Soc. Perkin Trans. II*, 2094
102. van Koningsveld, H. (1972). *Acta Crystallogr.*, **B28**, 1189
103. Johnson, P. L., Schaefer, J. P., James, V. J. and McConnel, J. F. (1972). *Tetrahedron*, **28**, 2901
104. Johnson, P. L., Cheer, C. J., Schaefer, J. P., James, V. J. and Moor, F. H. (1972). *Tetrahedron*, **28**, 2893
105. James, V. J. (1973). *Cryst. Struct. Commun.*, **2**, 205
106. de Graaff, R. A. G., Thiesen, M. Th., Rutten, E. W. M. and Romers, C. (1972). *Acta Crystallogr.*, **B28**, 1576
107. Krebs, B., Pohl, S. and Glemser, O. (1972). *J. Chem. Soc. Chem. Commun.*, 548

108. Choi, C. S. and Prince, E. (1972). *Acta Crystallogr.*, **B28**, 2857
109. Valle, G., Del Pra, A. and Mammi, M. (1973). *Cryst. Struct. Commun.*, **2**, 169
110. Carrell, H. L. and Donohue, J. (1972). *Acta Crystallogr.*, **B28**, 1566
111. Haasnoot, J. G., Verschoor, G. C., Romers, C. and Groeneveld, W. L. (1972). *Acta Crystallogr.*, **B28**, 2070
112. Gerdil, R. (1972). *Helv. Chim. Acta*, **55**, 2159
113. Clark, G. R. and Palenik, G. J. (1973). *J. Chem. Soc. Perkin Trans. II*, 194
114. Mackenzie, R. K., MacNicol, D. D., Mills, H. H., Raphael, R. A., Wilson, F. B. and Zabriewicz, J. A. (1972). *J. Chem. Soc. Perkin Trans. II*, 1632
115. Bassi, I. W., Scordamaglia, R. and Fiore, L. (1972). *J. Chem. Soc. Perkin Trans. II*, 1726
116. Sakabe, N., Sakabe, K., Ozeki-Minakata, K. and Tanaka, J. (1972). *Acta Crystallogr.*, **B28**, 3441
117. Macintyre, W. M. and Tench, A. H. (1973). *J. Org. Chem.*, **38**, 130
118. Abrahamsson, S. and Rehnberg, G. (1972). *Acta Chem. Scand.*, **26**, 3309
119. Naganathan, P. S. and Venkatesan, K. (1972). *Acta Crystallogr.*, **B28**, 552
120. Edsall, J. T., Flory, P. J., Kendrew, J. C., Liquori, A. M., Nemethy, G., Ramachandran, G. N. and Scheraga, H. A. (1966). *J. Mol. Biol.*, **15**, 399
121. Wei, C. H., Doherty, D. G. and Einstein, J. R. (1972). *Acta Crystallogr.*, **B28**, 907
122. Hewlins, M. J. E. (1972). *J. Chem. Soc. Perkin Trans. II*, 275
123. Flippen, J. L., Gilardi, R. D. and Karle, I. L. (1972). *Acta Crystallogr.*, **B28**, 360
124. Hvoslef, J. (1972). *Acta Crystallogr.*, **B28**, 916
125. Portheine, J. C., Romers, C. and Rutten, E. W. M. (1972). *Acta Crystallogr.*, **B28**, 849
126. Weeks, C. M., Cooper, A. and Norton, D. A. (1971). *Acta Crystallogr.*, **B27**, 531
127. Mez, H.-C. and Rihs, G. (1972). *Helv. Chim. Acta*, **55**, 375
128. Thom, E. and Christensen, A. T. (1971). *Acta Crystallogr.*, **B27**, 794
129. Gilardi, R. D., Karle, I. L. and Karle, J. (1972). *Acta Crystallogr.*, **B28**, 2605
130. Hamanaka, T., Mitsui, T., Ashida, T. and Kakudo, M. (1972). *Acta Crystallogr.*, **B28**, 214
131. Camerman, A. and Camerman, N. (1971). *Acta Crystallogr.*, **B27**, 2205
132. Camerman, A. and Camerman, N. (1972). *J. Amer. Chem. Soc.*, **94**, 268
133. Mathew, M. and Palenik, G. J. (1971). *J. Amer. Chem. Soc.*, **93**, 497
134. Bergin, R. (1971). *Acta Crystallogr.*, **B27**, 381
135. Bergin, R. (1971). *Acta Crystallogr.*, **B27**, 2139
136. Bergin, R. and Carlström, D. (1971). *Acta Crystallogr.*, **B27**, 2146
137. Beale, J. P. (1972). *Cryst. Struct. Commun.*, **1**, 223
138. Beale, J. P. (1972). *Cryst. Struct. Commun.*, **1**, 67
139. Beale, J. P. (1972). *Cryst. Struct. Commun.*, **1**, 297
140. Beale, J. P. and Grainger, C. T. (1972). *Cryst. Struct. Commun.*, **1**, 71
141. Carlström, D. (1973). *Acta Crystallogr.*, **B29**, 161
142. Hearn, R. A. and Bugg, C. E. (1972). *Acta Crystallogr.*, **B28**, 3662
143. Griffith, E. A. H. and Robertson, B. E. (1972). *Acta Crystallogr.*, **B28**, 3377
144. Tsoucaris, D., de Rango, C., Tsoucaris, G., Zelwer, Ch., Parthasarathy, R. and Cole, F. E. (1973). *Cryst. Struct. Commn.*, **2**, 193
145. Clark, G. R. and Palenik, G. J. (1972). *J. Amer. Chem. Soc.*, **94**, 4005
146. Pokrywiecki, S., Weeks, C. M. and Duax, W. L. (1973). *Cryst. Struct. Commun.*, **2**, 63
147. Pokrywiecki, S., Weeks, C. M. and Duax, W. L. (1973). *Cryst. Struct. Commun.*, **2**, 67
148. de Camp, W. H. and Ahmed, F. R. (1972). *Acta Crystallogr.*, **B28**, 3484
149. Ahmed, F. R. and de Camp W. H. (1972). *Acta Crystallogr.*, **B28**, 3489
150. DeLacy, T. P. and Kennard, C. H. L. (1972). *J. Chem. Soc. Perkin Trans. II*, 2148
151. DeLacy, T. P. and Kennard, C. H. L. (1972). *J. Chem. Soc. Perkin Trans. II*, 2141
152. Holan, G., Kowala, C. and Wunderlich, J. A. (1973). *J. Chem. Soc. Chem. Commun.*, 34
153. DeLacy, T. P. and Kennard, C. H. L. (1972). *J. Chem. Soc. Perkin Trans. II*, 2153

5
Natural Products—Structure Determination

E. HASLAM
University of Sheffield

5.1	INTRODUCTION AND GENERAL OBSERVATIONS		125
5.2	DEVELOPMENTS IN STRUCTURAL METHODS		127
	5.2.1	^{13}C *Nuclear magnetic resonance*	128
	5.2.2	*Chemical ionisation mass spectrometry*	132
5.3	CONFIGURATION AND CONFORMATION		137
	5.3.1	*Abscisic acid*	137
	5.3.2	*Some natural γ-lactones*	140
	5.3.3	*Bacterial dihydrodiols*	141
5.4	NEW GROUPS OF NATURAL PRODUCTS		142
	5.4.1	*Antibiotics*	142
		5.4.1.1 *Cephamycins*	142
		5.4.1.2 *Antibiotics in ion transport*	145
		5.4.1.3 *Cytochalasans*	148
	5.4.2	*Marine natural products*	152
	5.4.3	*Miscellaneous new structures*	156

5.1 INTRODUCTION AND GENERAL OBSERVATIONS

The last fifty years of the nineteenth century marked a period of remarkable progress and activity in organic chemistry. Some measure of this progress can be judged from the various syntheses of natural products carried out at the turn of the century, e.g. the synthesis of (±)-α-terpineol completed by W. H. Perkin[1]. Prior to his description of the synthesis, Perkin remarked 'this investigation was undertaken with the object of synthesising terpin, terpineol and dipentene not only on account of the interest which always attaches to syntheses of this kind but also in the hope that a method of synthesis might be devised of such a simple kind that there would no longer be room for doubt as to the constitution of these important substances'. Writing in this vein, Perkin clearly enunciated one of the traditional roles of organic synthesis, namely the verification of structure. However, because perhaps this function is to some degree an arbitrary one, its use has declined in recent years as a corresponding increased reliance has been placed on spectroscopic methods of structure determination.

Writing fifty years after Perkin's synthesis of (±)-α-terpineol in an essay entitled 'Synthesis', one of its grandmasters, R. B. Woodward[2], noted with some perception 'the present and growing power of x-ray diffraction methods for ascertaining structures'. Whilst at that time few structures had been established solely by x-ray methods, Woodward nevertheless foresaw that these developments would 'revive, for some time at least, the function of synthesis in dealing the *coup de grace* to problems of structure'. The most casual observer of the natural products chemistry scene over the past decade could scarcely have failed to notice how the first of Woodward's observations has been steadily borne out. The literature reveals a rapidly growing dependence on x-ray crystallographic analysis as a means of natural product structure determination. Thus, for example, in volume 95 (1973) of the *Journal of the American Chemical Society* in structural reports on 21 new natural products, x-ray crystallographic analysis was utilised either wholly or in part to determine structural or stereochemical details in 15 of these. This trend is even more clearly defined in journals devoted to short communications. Less clearly portrayed, however, is the role which the art of synthesis still continues to play in natural product structure determination. Synthesis indeed often provides not only the simplest but frequently the most rigorous proof of structure, and amongst a range of topics and developments reviewed in this chapter this will provide a recurrent and at times dominant theme.

Two recent alkaloid syntheses by Wuonola and Woodward[3] leading to revisions of previous structural proposals provide a most appropriate starting point. In 1969 Arndt, Eggers and Jordaan[4] reported the isolation of several alkaloids from *Macrorungia longistrobus* and structures (1) and (2) were

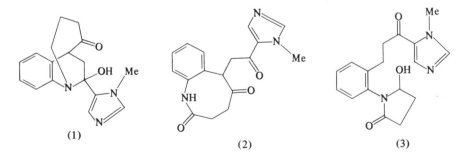

(1)   (2)   (3)

assigned, respectively, to isolongistrobene and dehydrolongistrobene on the basis of spectroscopic and chemical data. Reinterpretation of these data,

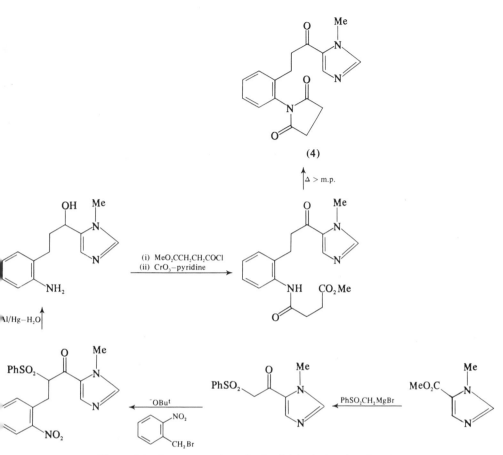

Figure 5.1 Structure and synthesis of dehydrolongistrobene[3]

however, led to the proposal by Wuonola and Woodward[3] of the alternative structures (3) and (4) for the two alkaloids and these suggestions were decisively supported by rational syntheses of both alkaloids. The synthesis of dehydro-longistrobene (4) is illustrated in Figure 5.1.

5.2 DEVELOPMENTS IN STRUCTURAL METHODS

Developments in two important physical methods as they relate to natural products are discussed below. Some remarks by Sir Robert Robinson on the state of the art of chemistry are however perhaps appropriate both as an introduction and to sound a cautionary note. In a recent interview[5] he said 'So the use of more and more sophisticated physical methods will make it more and more easy to obtain the structure; this is a very good thing because it saves a very great deal of time, but it does not mean that the subject has been exhausted—*you have still got to go back and work on the chemistry of that compound*' (author's italics).

5.2.1 ^{13}C Nuclear magnetic resonance

The impact on organic chemistry which many envisaged this spectroscopic technique would have is now beginning to be felt over a wide area. The period of gestation for a new physical method is now remarkably small and although spectral compilations continue to be published and empirical rules quantified and evaluated, this spectroscopic technique has been employed in the past two years to considerable advantage in structural work. Undoubtedly its greatest contributions have been in the area of biogenesis, where the use of ^{13}C as a tracer has been shown to be of considerable advantage in a number of examples. Until the advent of ^{13}C tracer techniques, verification of bio-synthetic processes was most commonly achieved by use of ^{14}C tracer studies. Determination of the specific activity of a given carbon atom in a metabolite using the ^{14}C method normally involves a degradative scheme that selectively isolates a particular carbon atom in a molecular fragment— a procedure which may be difficult and is often time consuming. Detection and identification of the labelled sites using ^{13}C as a tracer can usually be accomplished by either ^{1}H (satellite method) or ^{13}C n.m.r. spectroscopy and without chemical degradation.

Several examples of these procedures have been published recently[6-13], but some of the most elegant examples of the applications of this technique concern its use in determining the overall molecular features of biochemical rearrangements and details of the stereochemistry of some biological pro-cesses. Thus Battersby and his collaborators[14,15] have used the ^{13}C tracer to unravel the complex rearrangement process in the biosynthesis of type III porphyrins and other groups have examined stereochemical features of the biosynthesis of penicillins and cephalosporins[16, 17]. An example of its use in biogenetic studies which illustrates the type of information which it is now possible to provide using the ^{13}C n.m.r. technique concerns[18] the biosynthesis of vitamin B_{12} (5). Experiments with whole cells of *Propionibacterium shermanii* have implicated uroporphyrinogen III (7) as a precursor for Vitamin B_{12}. Seven of the methyl groups of the vitamin have been shown to arise from the methyl group of L-methionine (8) and Shemin and Bray[19] demonstrated that only one of the methyl groups at C-12 was derived from (8); the other was formed by decarboxylation of the acetic acid side chain of the precursor (7). ^{13}C-labelled (methyl group) L-methionine (8) gave vitamin B_{12} (5) enriched with ^{13}C and hydrolysis (trifluoroacetic acid) produced cobinamide (6a) and neocobinamide (6b; part structure) in which epimerisa-tion had occurred at C-13 in the ring c. The conformation of the c ring of cobinamide[20, 21] places the α-methyl group *syn* periplanar to the adjacent axially oriented propionamide side chain at C-13. Such a juxtaposition of groups would be expected to produce a γ-effect on the ^{13}C chemical shift of the α-methyl group[22]. Conversely, in neocobinamide (6b) this effect is absent. Examination of the ^{13}C Fourier transform spectrum of (6a) showed seven methyl resonances 20–27 p.p.m. downfield from TMS. In neocobinamide a downfield shift of one methyl group from 23.8 to 35.5 p.p.m. was observed and it was therefore concluded that L-methionine inserts one methyl group at the α-position at C-12 at a late stage of the biosynthetic process (Figure 5.2).

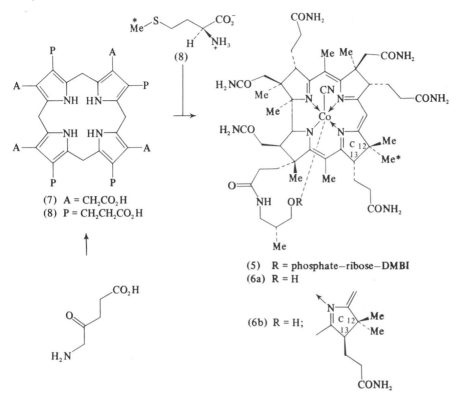

(7) A = CH_2CO_2H
(8) P = $CH_2CH_2CO_2H$

(5) R = phosphate–ribose–DMBI
(6a) R = H

(6b) R = H;

Figure 5.2 Biosynthesis of the C-12 α methyl group in cyanocobalamin (5)[18]

Compilations of ^{13}C n.m.r. data still continue to pour out and, amongst others, publications concerning penicillins[23], aminoglycosides[24], dextrans[25], purines[26] and pyrimidines[27], steroids[28], pteridines[29], rifamycins[8], strepto-varicins[9], alkaloids[30] and corrins[31] have appeared. In the nucleotide field, use of ^{13}C n.m.r. has also been made to explore conformational problems[32, 33]. An increasing number of publications describing the ^{13}C n.m.r. spectra of carbohydrates have also appeared[34-40] and these have led to some interesting empirical observations. Thus it has been shown that the chemical shift of the anomeric carbon atom is strongly dependent on the anomeric configuration and measurements of the $^{13}C-^1H$ coupling constants in carbohydrates have shown that the $^{13}C-^1H$ coupling constant at the anomeric centre is significantly larger than the other geminal C–H coupling constants involving C-2 to C-6. The latter have values of 143–145 Hz in free sugars and methyl glycosides and do not appear to be very dependent on stereochemistry. The

^{13}C–^{1}H values at C-1, on the other hand, show a clear dependance on the orientation of the substituents at C-1. Thus in the methyl glycosides in which H-1 is axially oriented, $J = $ 158–162 Hz, whereas with H-1 in an equatorial position a higher coupling constant (169–171 Hz) is observed. Typical examples are the glycosides (9) and (10) and these observations will clearly be of considerable interest in later structural work.

(9) $J(^{13}C-^{1}H) = 160.0\,Hz$ (10) $J(^{13}C-^{1}H) = 169.5\,Hz$

Applications of ^{13}C n.m.r. spectroscopy to determine the structure of natural products *per se*, however, remain few in number. Nakanishi and his collaborators[41] have used the technique to elucidate the structures of tingenin A and tingenin B—two triterpenes related to the potent but toxic anti-tumour agent pristimerin—and Wenkert[42] has used ^{13}C n.m.r. spectroscopy to define

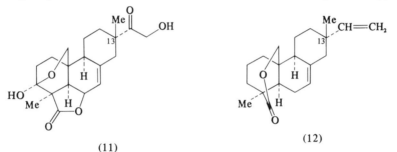

(11)   (12)

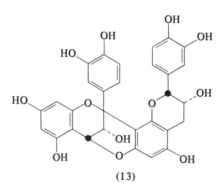

(13)

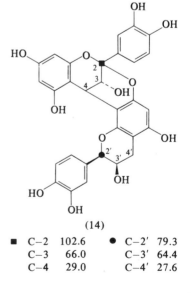

(14)

■ C–2	102.6	● C–2′	79.3
C–3	66.0	C–3′	64.4
C–4	29.0	C–4′	27.6

the stereochemistry at C-13 in the diterpene annonalide[43, 44] (11). This was accomplished by [13]C n.m.r. measurements on the lactone (12)—a degradation product of the natural product—and comparison with the [13]C n.m.r. chemical shifts in related model compounds.

A similar approach based on an analysis of chemical shifts was used independently by two groups[45, 46] to establish the structure of proanthocyanidin A-2 (14), a phenolic metabolite of *Aesculus hippocastanum*. Other spectroscopic data pointed to the two alternative structures (13) and (14) for the natural product and whilst structure (13) contains two carbon atoms

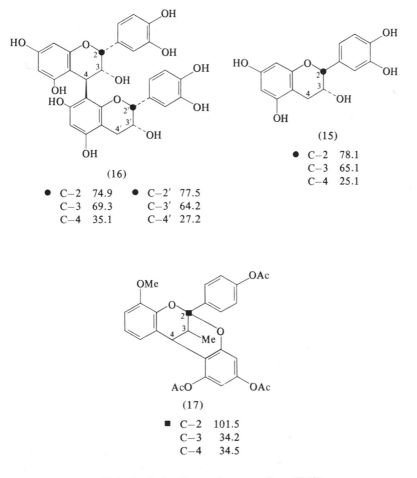

(16)

- C–2 74.9   ● C–2′ 77.5
  C–3 69.3   C–3′ 64.2
  C–4 35.1   C–4′ 27.2

(15)

- C–2 78.1
  C–3 65.1
  C–4 25.1

(17)

■ C–2 101.5
  C–3   34.2
  C–4   34.5

(^{13}C chemical shifts are in p.p.m. from TMS)

with very similar chemical environments to C-2 in the model (−)-epicatechin (15; ●), the structure (14) contains only one such atom and a carbon atom ■) with the distinctive characteristics of an acetal carbon. Analysis by [13]C n.m.r. spectroscopy and utilising the models (15)–(17) confirmed the alternative (14) as the correct one.

5.2.2 Chemical ionisation mass spectrometry

The potential of chemical ionisation mass spectrometry[47] as an analytical and structural tool within the field of complex organic natural products has still to be explored in any detail, although some studies towards this end have been made.

Chemical ionisation (CI) mass spectrometry is a relatively new technique whereby gaseous molecules are ionised by collision with a chosen set of reagent ions rather than by electron impact (EI). Field and Munson[48, 49] pioneered this development in 1966. They recognised that the ion–molecule reactions in methane at relatively high pressure (0.5–1.0 Torr) were very sensitive to trace quantities of gaseous impurities and mass spectra were generated characteristic of the extraneous material. In CI mass spectrometry, ions of a selected reagent gas (hydrogen, methane, isobutane) are formed by a combination of electron impact and ion–molecule reactions. Ionisation of the sample molecules, which are present at low concentrations (10^{-4}–10^{-5} Torr), then takes place by reaction with these ions. This may result in ion–molecule transfer of charged entities, such as protons and hydride ions, or by ion association. Some of the primary product ions may possess sufficient excess internal energy to allow fragmentation to ions of lower mass. Nevertheless the CI mass spectra tend to be much simpler in appearance than the corresponding EI spectra due to several factors such as the generally low excess internal energies in chemical ionisation processes and the even electron character of the primary product ions. Thus chemical ionisation spectra often possess significant abundances of ions in the region of the molecular weight of the additive (usually the protonated molecular ion, MH^+). The fragmentation patterns often yield structural evidence of a different nature from that obtained by electron impact mass spectrometry.

Clark-Lewis and his collaborators[50] and Kingston and Fales[51] have both examined in some detail the chemical ionisation mass spectra of a range of flavonoid compounds. The application of electron impact mass spectrometry to this group of naturally occurring oxygen heterocyclic compounds has previously been examined and discussed in some detail[52, 53] and a strict comparison of the two techniques has therefore been possible. Kingston and Fales[51] thus showed that with a number of flavones, flavonols and their methyl ethers, examination by chemical ionisation mass spectrometry with methane as reactant gas gave very little fragmentation. There was little advantage to be gained therefore in the use of chemical ionisation mass spectrometry for their structure elucidation. In contrast, the flavanones and 3-hydroxyflavanones which were examined did show significant diagnostic fragmentations. These pathways of fragmentation were most satisfactorily rationalised in terms of the concept of localised proton transfer. Most of the fragmentations appeared to arise from isomeric progenitors protonated at different sites in the heterocyclic ring of the flavonoid molecule. Clark-Lewis and his co-workers came to very similar conclusions.

Thus the flavanones naringenin (18) and hesperitin (19) showed significant peaks corresponding to the fragmentation patterns shown (Figure 5.3). Protonation on the heterocyclic oxygen atom leads to a protonated species

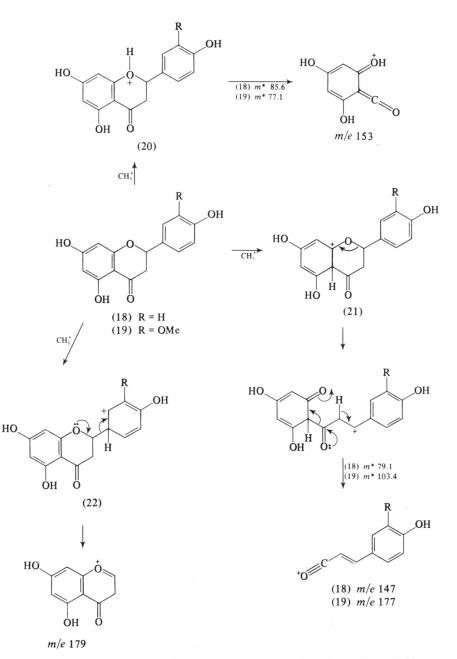

(18) m* 85.6
(19) m* 77.1

m/e 153

(20)

(18) R = H
(19) R = OMe

(21)

(22)

(18) m* 79.1
(19) m* 103.4

(18) m/e 147
(19) m/e 177

m/e 179

Figure 5.3 Chemical ionisation mass spectrometry of naringenin and hesperitin[51]

(20) which then decomposes via a retro-Diels–Alder reaction to give the fragment ion m/e 153. On the other hand, protonation on the A ring (21)— to give a resonance stabilised ion—could then lead to fragmentation in the alternative manner shown. The intense ion at m/e 179 in the spectra of both flavanones was plausibly rationalised in terms of the loss of ring C from both natural products via the protonated species (22).

Mitscher, Showalter and Foltz[54] have similarly applied the technique of chemical ionisation mass spectrometry to a number of clinically important macrolide antibiotics. The authors claim considerable promise for the method in this field and noted particular advantages when compared with electron impact mass spectrometry. The EI mass spectra of the macrolide antibiotics are very complex with nearly all of the abundant ions in the low mass region, whilst the CI mass spectra, using isobutane as the reactant gas, show relatively simple fragmentation patterns which provide important structural information. The spectra normally show prominent protonated molecular ion + 1 peaks (MH^+) and peaks due to cleavage of the glycosidic bonds. Most of the remaining prominent fragment ions may be rationalised on the basis of the loss of neutral oxygenated molecules such as water, methanol and acetic acid. However, a few important fragmentations were noted involving the cleavage of C—C bonds which are structurally diagnostic. In the CI mass spectrum of erythromycin B (27) the base peak is at m/e 718 (MH^+) and peaks at m/e 756 and m/e 774 were attributed to adduct ions ($M^+ + C_3H_3$ and $M^+ + C_4H_9$) which are typically observed in CI mass spectra using isobutane. The direct loss of cladinose from MH^+, with or without its glycosidic oxygen, gave peaks at m/e 542 and 560, respectively, and ions due to this sugar in the low mass region of the spectrum were attributed as shown (24), (25) and

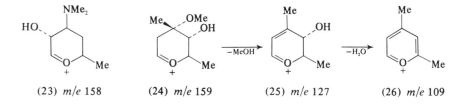

(23) m/e 158   (24) m/e 159   (25) m/e 127   (26) m/e 109

(26). The presence of the desosamine structural unit was indicated by the peak (23; m/e 158).

An interesting correlation of the C—C bond fragmentations in the CI mass spectrum of erythromycin B (27) was made with established solution chemistry in this field. Thus the abundant $MH^+ — H_2O$ ion (m/e 700) was attributed in the manner shown (Figure 5.4; 28) to a reaction sequence which is analogous to the known acid-catalysed reaction which occurs in solution in those macrolides where the hydroxyl group at C-6 is conformationally juxtaposed for attack of the C-9 carbonyl group. The intense peak at m/e 99 was similarly rationalised as resulting from protonation of the C-9 ketone, ring opening of the lactone and cleavage of the C-10—C-11 bond by a retro-aldol fragmentation (Figure 5.4; 29). The expected mass shifts for these fragmentations were observed in the CI mass spectra of other macrolides.

Several cases have been noted in which stereochemical differences in

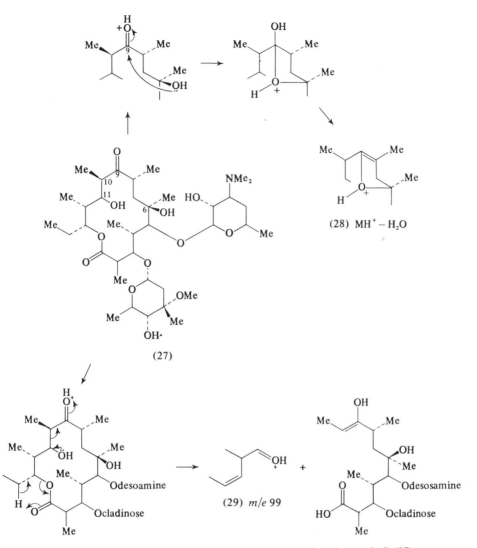

Figure 5.4 Chemical ionisation mass spectrometry of erythromycin B (27)

natural product structures are manifest in the chemical ionisation mass spectra. Typical examples are various alkaloids[55], steroids[56] and esters of maleic and fumaric acid[57]. Fales and his co-workers[58] have also utilised the technique as a stereochemical and conformational probe in studies of epimeric steroidal amino alcohols. The results were in good agreement with similar measurements obtained by other means, such as infrared spectroscopy. The chemical ionisation mass spectra of several steroidal amino alcohols (ring A substituted 1,2 and 1,3 disposition of the two functional groups) were studied using isobutane as a reagent gas. It was observed[58] that the loss of water from the MH$^+$ ion (protonated at the hydroxyl group) only occurred when the distance between the oxygen and nitrogen atoms was too large to allow the formation of a hydrogen bond.

Chemical ionisation mass spectrometry was used for the first time as a tool in organic structure determination by Arsenault, Althaus and Divekar[59] in the determination of structure of the antibiotic botryodiplodin from *Botryo-diplodia theobromae*. The metabolite is an anti-leukaemic agent[60] and two to three hours after its application turns the skin of individuals varying shades of pink. The electron impact mass spectrum of botryodiplodin was not a reliable source of information because of the variation of the spectrum obtained under different operating conditions. The chemical ionisation mass spectrum of the antibiotic, however, led to a firm assignment of the molecular weight as 144 and an elemental composition of $C_7H_{12}O_3$. Combined with chemical and other spectroscopic data this led to the gross structure of 2-hydroxy-3-methyl-4-acetyltetrahydrofuran (30) for the antibiotic. The relative stereochemistry of the groups around the tetrahydrofuran

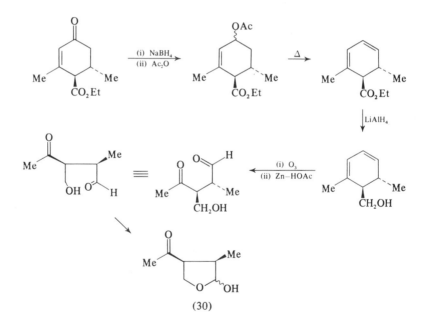

Figure 5.5 Synthesis of *dl*-botryodiplodin[62]. (Formulae show relative stereochemistry.)

ring was not readily determined, however, because of the ambiguity of assignments based on coupling constants in tetrahydrofuran and its derivatives[61]. However, a *cis* relationship between the *C*-methyl and *C*-acetyl groups was suggested. Proof of the relative stereochemistry in botryodiplodin has been obtained by synthesis[62] of the *dl*-form (Figure 5.5).

5.3 CONFIGURATION AND CONFORMATION

The physical and theoretical principles of many of the chiroptical methods which may be employed to elucidate the chirality, relative configuration or conformational properties of an organic natural product are not well understood. Empirical rules have usually been devised based on the extensive correlations which are possible within a given set of experimental data. Many of the rules thus devised have enjoyed particular success in their subsequent application, but occasionally wrong assignments result and this normally leads to a re-examination of the extent of applicability of the rule in question. Some examples in the recent literature of natural product chemistry where this has occurred are, amongst other configurational and conformational investigations, outlined below.

5.3.1 Abscisic acid

Abscisic acid is an unusual sesquiterpene plant growth inhibitor which promotes senescence and abscission of leaves and induces dormancy in buds and seeds. Addicott and his collaborators[63] first isolated abscisin II from young cotton fruit and this was later shown to be identical with the crystalline dormin from sycamore (*Acer pseudoplatanus*)[64]. The structure of abscisin II—later named abscisic acid—was proposed as (31) and confirmed by a synthesis of the racemic form by Cornforth, Milborrow and Ryback[65]. The structure of abscisic acid (31) contains one centre of asymmetry and the natural isomer is dextrorotary and exhibits a very intense positive Cotton effect with extrema at 287 and 245 nm. This property has been used in the identification and estimation of the plant growth regulator in vegetative tissues[66]. The absolute stereochemistry around its one centre of asymmetry was deduced[67] by comparing the molecular rotation values of the two diol esters, epimeric at C-4', obtained by borohydride reduction of its methyl ester. Mills' empirical rule[68] for allylic cyclohexenols was then used to assign the absolute configuration at C-4' and by relating the stereochemistry at C-4'

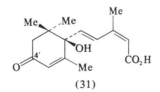

(31)

to that at C-1' the absolute stereochemistry of natural abscisic acid followed as (31).

Several lines of enquiry have suggested, however, that this assignment was incorrect. Thus Burden and Taylor[69] showed that photo-oxidation of violaxanthin[70] gave xanthoxin [a mixture of the stereoisomers (32) and (33)]. Chromium trioxide–pyridine treatment of xanthoxin gave a mixture of abscisic aldehyde (34) and *trans*-abscisic aldehyde (35). Further oxidation

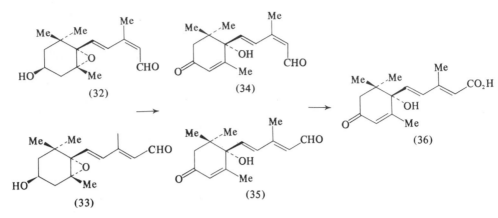

(32)   (34)   (36)

(33)   (35)

of this mixture yielded predominantly *trans*-(+)-abscisic acid (36), a geometrical isomer of the natural growth inhibitor. The acid had an o.r.d. curve
almost identical with the natural abscisic acid and this conversion indicated
that the assignment of stereochemistry to either violaxanthin or abscisic
acid (31) was incorrect. Related work in other laboratories[71,72] implied
similarly that a revision of the absolute stereochemistry of abscisic acid
(31) was required.

Three groups have independently demonstrated that the absolute configuration of natural (+)-abscisic acid is *S*. Harada[73] applied the exciton
chirality method to *trans*-abscisic acid and demonstrated that it has the
S-configuration and hence that the natural (+)-*cis*-abscisic acid has also this
configuration (37). Ryback[74] provided unambiguous proof of this result by

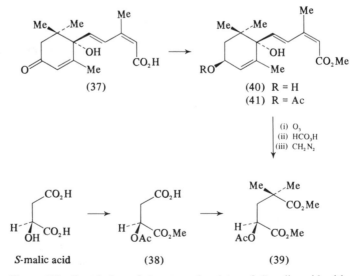

(37)   (40) R = H
         (41) R = Ac

(i) O₃
(ii) HCO₃H
(iii) CH₂N₂

S-malic acid   (38)   (39)

Figure 5.6 Correlation of the stereochemistry of *S*-malic acid with
cis-(+)-abscisic acid (37)[74]

a chemical correlation with *S*-malic acid (Figure 5.6). Methyl (*S*)-2-acetoxy-3-carboxypropionate (38), derived from *S*-malic acid[75], was electrolysed in methanol with an excess of methyl dimethylmalonate to give the ester (39) in its laevorotatory form. The same ester was also obtained from the (+)-*trans*-diol ester (40) derived from (+)-*cis*-abscisic acid. Acetylation gave the monoacetate (41) which was transformed to (39) by successive ozonolysis, oxidation with performic acid, and methylation with diazomethane. This observation therefore defines the absolute stereochemistry of the *trans*-diol acetate as (41) and hence that of natural (+)-*cis*-abscisic acid as (37).

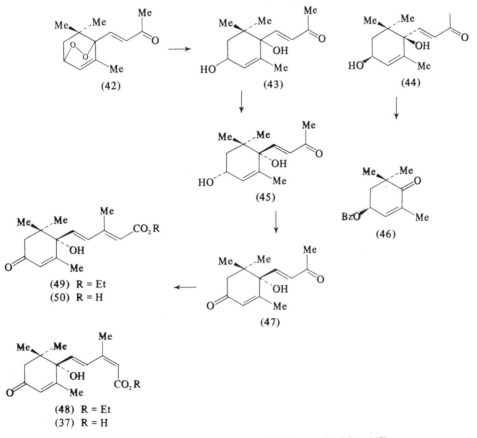

Figure 5.7 Synthesis of (+)-*cis*- and (+)-*trans*-abscisic acid[76]

Koreeda, Weiss and Nakanishi[76] have provided an alternative proof of the stereochemistry of (+)-*cis*-abscisic acid by a synthesis of the acid (and its enantiomeric form), using a chemical approach with the exciton chirality method to elucidate configuration (Figure 5.7). Hydrogenation (Lindlar) of the peroxide (42) gave the *dl-cis*-diol (43), which was separated into its diastereoisomeric forms (44) and (45) by liquid chromatography of the (+)-α-methoxy-α-trifluoromethylphenyl acetate esters. The absolute configuration of the less polar diol (44) was deduced from the c.d. spectra of the degradation product (46), which showed a split Cotton effect due to inter-action between the benzoate and enone chromophores. The other diol (45) was oxidised to the bis-enone (47), which was converted into the mixture of *cis*- and *trans*-esters (48) and (49) by a Wittig reaction. These were separated and hydrolysed to give respectively the natural (+)-*cis*-abscisic acid (37) and the corresponding (+)-*trans*-abscisic acid (50).

In the earlier work the configuration (31) was specified as *S* according to the then suggested system of nomenclature[77]. However, a revision of these rules in 1966[78], in which a modified procedure for dealing with double bonds was introduced, led to a change in the sequence rules whereby (31) became *R*. By a fortunate coincidence the absolute configuration of natural (+)-*cis*-abscisic acid as specified by formula (37) is therefore *S* and no correction of the earlier literature on this particular point of nomenclature is necessary.

5.3.2 Some natural γ-lactones

Comparable stereochemical ambiguities have arisen recently with some naturally occurring lactones. Thus application[79] of an empirical rule[80] relating the sign of the lactone Cotton effect to the xanthanolides parthe-mollin (51), apachin (52) and ivambrin (53) indicated a *trans* fusion of the

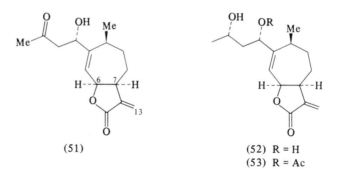

(51)

(52) R = H
(53) R = Ac

lactone ring. On the other hand, use of the modified Hudson–Klyne rule for lactones[81] suggested[79] that parthemollin (51) and the compounds corre-lated with it should be considered as *cis*-lactones, and this assignment was supported by the coupling constant $J_{7,13}$ of 3 Hz in (51) and its derivatives. This problem was resolved by an x-ray diffraction study[82] of parthemollin (51), which showed it to be a *cis*-lactone. [1]H n.m.r. and c.d. studies permit a correlation of the stereochemistry of parthemollin with ivalbatin and other closely related xantholides[83].

The full stereochemical details of the structure of two new sesquiterpene lactones carolenalin (54) and carolenin (55) from *Helenium autumnale* were also deduced directly by the x-ray method[84]. The structure and some facets of the stereochemistry of these guaianolides were deduced[85] from the usual spectroscopic data, although the orientation of the C-11 methyl was not determined. The assignment of the stereochemistry at the two ring junctions

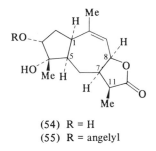

(54) R = H
(55) R = angelyl

depended heavily on the value of the $J_{7,8}$ proton coupling constant of 7.2 Hz and the result of NOE studies which produced an enhancement of the H-8 signal when H-1 was irradiated. These data were interpreted as being indicative of the presence of a *trans*-fused lactone. They are also compatible, however, with a *cis*-fused geometry in which the central seven-membered ring adopts a boat type conformation with H-1 and H-8 both axially disposed. X-Ray analysis of carolenalin acetate showed indeed that both ring fusions were *cis* but that the seven-membered ring adopts a flattened chair conformation.

5.3.3 Bacterial dihydrodiols

Both mammals and bacteria oxidise aromatic hydrocarbons to dihydrodiols. This pathway of metabolism occurs in mammals via the formation of an arene oxide which is subsequently hydrolysed (catalysed by epoxide hydrase) to yield a *trans*-dihydrodiol. In direct contrast, the dihydrodiols formed during the bacterial degradation of aromatic hydrocarbons have a *cis* stereochemistry[86] and both atoms of oxygen in the diol may be shown to originate from molecular oxygen. The absolute stereochemistry of the *cis*-dihydrodiol from naphthalene has been determined[87] as (56) and recent work, both chemical[88] and x-ray crystallographic, has shown the diol derived from toluene to have the absolute stereochemistry as shown in (57).

A mutant strain of *Pseudomonas putida* which lacked the ability to convert the dihydrodiol once it was metabolised to the corresponding catechol derivative was used to convert toluene and its derivatives to the dihydrodiol[88]. An x-ray analysis[89] of the adduct (59) derived from the diol (57) by reaction with the triazoline dione (58) gave the absolute configuration of (57) directly. In an alternative method[88] the same configuration was deduced by correlation with substances of known stereochemistry. Hydrogenation of (57) gave the diols (60) and (61), which were separated as their C-1 monobenzoates. The assignment of relative stereochemistry in (60) and (61) was deduced by a synthesis of the *dl*-form of (61) from 3-methylcyclohexene and osmium tetroxide. The formation of the diastereoisomer (61) in this reaction is rationalised on the basis of steric approach control and attack of the reagent from the least hindered side of the molecule. Oxidation of the diol (60) gave the known (—)-(2R)-2-methyladipic acid (62) and hence the configuration of (60) followed as 1S,2R,3R. The dibenzoate chirality rule of Nakanishi and Harada[90] was also used to confirm this assignment and hence that of the

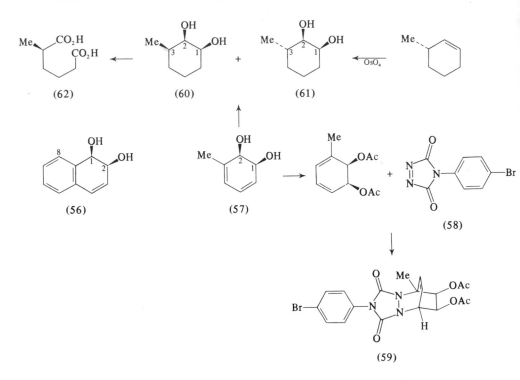

(62)   (60)   (61)

(56)   (57)   (58)

(59)

metabolite (57) as 1S,2R. Significantly, the configuration of the diol grouping in the toluene dihydrodiol (57) may be considered to be the same as in the naphthalene metabolite (56) if the methyl group and the C-8 methine are treated as analogous positions.

5.4 NEW GROUPS OF NATURAL PRODUCTS

5.4.1 Antibiotics

Intensive programmes both of a synthetic nature and those based on the screening of microbial metabolites continue to lead to important developments in the chemistry of antibiotics. Some of the recently discovered groups of antibiotics are discussed below.

5.4.1.1 Cephamycins

A serious limitation to the value of penicillins in medicine arose from the emergence of strains of bacteria that showed resistance to the more commonly used members of this group of antibiotics. This was due to the ability of the bacteria to produce the enzyme penicillinase which catalyses the hydrolytic ring opening of the β-lactam ring of a penicillin with the formation of an inactive penicilloic acid. Since the early 1950s a great deal of work has impinged on the problems posed by the penicillinase producing bacteria. Considerable research effort has, for example, been devoted towards obtaining new penicillins by variation of the side chain attached to the 6-amino group. In addition, the cephalosporins, a group of biogenetically related antibiotics, has also attracted wide interest because of their increased resistance towards β-lactamase enzymes. Abraham and his collaborators[91-93] first isolated cephalosporin C (64) from species of *Cephalosporium* and showed that although it was a powerful inducer of penicillinase activity in some

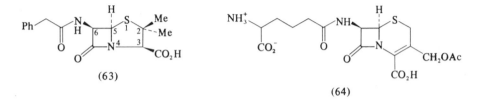

(63)

(64)

bacteria, it was hydrolysed some 5×10^3 times more slowly by the enzyme than penicillin G (63).

Latterly, interest in these two groups of antibiotics has been stimulated both by observations[94] on their probable mode of action in the inhibition of bacterial cell wall synthesis and by the discovery[95] of the cephamycins— naturally occurring 7α-methoxycephalosporins—which possess distinct antimicrobial properties. The cephamycins (65a–d) are produced by species of *Streptomyces* and, in addition to the presence of the unique 7α-methoxy group, they show some interesting variations in the nature of the side chain R including the occurrence of the biogenetically interesting α-methoxy-*p*-coumaroyl group (65c,d). The structure of the basic skeleton of the cephamycins was readily inferred from spectroscopic data. All possessed a band at 1770 cm^{-1} in the infrared which suggested the presence of a β-lactam carbonyl group; the characteristics of the 3-cephem chromophore[96] were readily distinguished in the ultraviolet spectra. High resolution mass spectra confirmed these deductions. The presence of the 7α-methoxy group was deduced from [1]H n.m.r. measurements. Thus in cephalosporin C (64) the H-6 proton occurs as a doublet, $\tau = 4.86$, $J_{6,7} = 4.7$ Hz ([²H₆]DMSO), but in (65a) the H-6 proton occurs as a singlet at $\tau = 4.84$, the signal for H-7 is absent and the spectrum shows an additional singlet (3-H) at $\tau = 6.47$ which was ascribed to the presence of a methoxyl group. The presence of the latter group was confirmed by alkoxyl analysis when the cephamycins (65a) and (65b) gave, respectively, 4.5% and 5.8% methoxyl.

Strominger and Tipper[94] had earlier postulated, on the basis of their model of penicillin action on bacterial cell wall synthesis, that 6α-methyl penicillins should have enhanced antimicrobial activity. This proposal and the discovery of the cephamycins has currently stimulated a great deal of interest and a large output of publications[97-104] concerning the synthesis of 6-substituted penicillins and 7-substituted cephalosporins. These synthetic routes have in addition lent support to the structural and stereochemical proposals previously made for the cephamycins. One route which has been utilised by

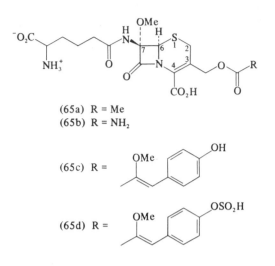

(65a) R = Me
(65b) R = NH$_2$

(65c) R =

(65d) R =

Christensen and his collaborators[97] to the cephamycins is shown in Figure 5.8. The synthesis of cephamycin (65a) by this procedure showed it to be identical with the natural product thus confirming the previous 7α-configurational assignment.

The formation of a single methoxyazide from either stereochemical form of the bromoazide (66) is presumed to occur by addition of methanol from the less hindered *exo* face of the α-oxo carbonium ion intermediate. Alternative procedures which have been developed for the synthesis of this type of compound depend analogously on steric approach control for selection of the correct stereochemistry at the centre undergoing substitution. Baldwin and his collaborators[102] have demonstrated that methanol adds to acylimines

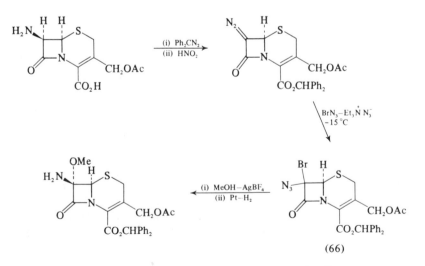

(66)

Figure 5.8 Synthesis of the cephamycin nucleus[97]

(67), derived from α-acetamido acids by a halogenation–dehydrohalogenation sequence, to afford methoxyamides and that the methanol adds stereoselectively to the α-face. *N*-Acylimine or carbonium ion intermediates have also been postulated by other workers as being involved in other methods for the introduction of the 7α-methoxy group into the penicillin or cephalosporin nucleus. Ratcliffe and Christensen[99] and Spitzer and Goodson[101] have thus utilised sequences in which the 7α-methylthio group is replaced by the

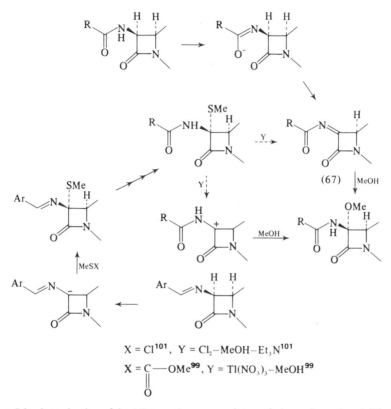

$$X = Cl^{101}, \quad Y = Cl_2-MeOH-Et_3N^{101}$$
$$X = \underset{\parallel}{\overset{}{C}}-OMe^{99}, \quad Y = Tl(NO_3)_3-MeOH^{99}$$

Figure 5.9 Introduction of the 6(7)α-methoxy group into cephalosporins and penicillins[98-104]

O-methyl group (Figure 5.9) and Koppel and Kochler[103] generate the acylimine intermediate directly with t-butyl hypochlorite from the corresponding amide anion.

5.4.1.2 Antibiotics in ion transport

Research in transport phenomena through biological membranes has developed along a very broad front in recent years and in one area these investigations have highlighted some intriguing relationships between the structure and function of antibiotics. Of particular interest from the point of view of membrane physiology is the fact that certain antibiotics such as the cyclododecapeptide valinomycin (68)[105] and the macrocyclic tetralactone nonactin (69)[106] produce striking effects on the transport of potassium but not sodium ions[107].

Eisenman[107] has summarised five possible mechanisms whereby ions might be transported across a biological membrane. One of these[108] is based on the

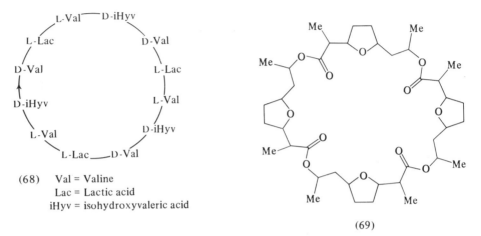

(68)   Val = Valine
        Lac = Lactic acid
        iHyv = isohydroxyvaleric acid

(69)

observation that various natural macrocyclic antibiotics [such as (68) and (69)] are able to induce ion permeation in mitochondria by the formation of lipid soluble charged complexes. More recently, structurally simpler synthetic polyethers[109, 110] have been shown to produce effects entirely comparable with those of the natural antibiotics. Spectroscopic[111] and x-ray crystallographic[112] studies of the potassium—nonactin complex show that the conformation of the 32-membered ring in the complex resembles the seam of a tennis ball with the carbonyl and ether oxygen atoms directed at the centre to the potassium ion and with the methyl substituents and methylene groups of the tetrahydrofuran rings on the exterior. The complex has approximate S_4 symmetry. ^1H n.m.r. data show that the nonactin has a very high affinity for potassium ions and that the potassium ion is bound without its water of hydration prior to entering the central aperture of the nonactin ring.

A new and rapidly growing group of antibiotics which also influence cation permeability in biological systems are the polyether polyalcohol monocarboxylic acids (mol. wt. 700–1000) typified by monensin (70)[113, 114], nigericin[115], X-537A[116], grisorixin[117], dianemycin[118], X-206[119], salinomycin[120] and the largest known member of this group, A-204A[121] (71; mol. wt. 945). Structure determinations within this group have been facilitated by spectroscopic measurements but have been heavily dependent on the use of x-ray crystallography. A number of different crystallographic structure determinations on antibiotics of this class coordinating the potassium ion or isomorphous to the potassium form have been made. Several common features are evident. Thus the carbon backbone of the antibiotic is folded around in a simple 'head to tail' loop which is fastened by a hydrogen bond

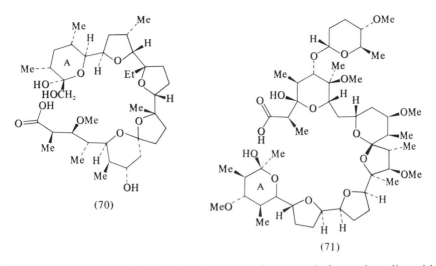

(70)

(71)

between one or more hydroxyl groups on ring A and the carboxylic acid group. The coordination to the central metal ion is variable but it is always to carboxylate, carbonyl, ether or alcohol oxygen atoms in the natural product. The atomic radius of the complexed ion also appears to be of some importance and measurements with monensin suggested a critical radius of 1.0 to 1.45 Å. C.d. measurements have also facilitated resolution of the solution conformational behaviour of a number of these antibiotics. Thus the antibiotic X-537A[122] exhibits strong optical activity in the presence of several amines and monovalent and divalent metals, but weak optical activity when free in aqueous ethanol. These changes in optical activity were correlated by Alpha and Brady with the presence of a disordered structure in aqueous ethanol and a cyclic 'head to tail' conformation in the presence of a complexing cation.

The cyclodecapeptide antamamide (72) from *Amanita phalloides* has in contrast a marked specificity for the sodium ion in complex formation. A three-dimensional structure was proposed[123] for the antamamide–sodium ion complex on the basis of spectroscopic and physicochemical studies. The complex has four intramolecular amide hydrogen bonds and the sodium ion is held in the inner cavity by ion–dipole interactions with the six remaining free amide carbonyl groups.

In the context of these studies it is important to note that the antibiotics of these various classes which have been examined have one property in common. They provide an appropriate environment for a cation of the proper valence and size and this is accomplished by the oxygen atoms of the antibiotic replacing the oxygen atoms of the water molecules which make up the hydration shell of the cation. The resulting complex has a hydrophobic exterior, a hydrophilic interior and a characteristic solubility in organic media which permits them to move into and through lipid membranes.

An unusual variation on this pattern is provided by the linear pentadecapeptide (gramicidin A; 73). Recent spectroscopic evidence[124] (u.v., c.d. and o.r.d., ^{1}H n.m.r.) suggests a helical conformation for this molecule. This

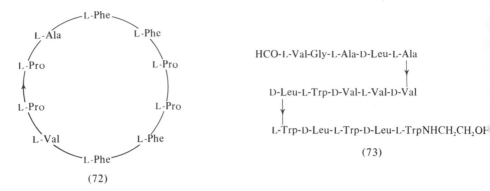

(72)

(73)

observation is of particular interest because of the capacity of the antibiotic to transport selectively cations across a lipid bilayer by the formation of trans-membrane channels. It has recently been proposed that the gramicidin A trans-membrane channel is formed by 'head-to-head' attachment of two helical conformations of the peptide and that the mechanism of transport is one in which the ion moves along the core of the helix by an ion-induced relaxation of the channel's helical conformation.

5.4.1.3 Cytochalasans

The cytochalasans are a group of microbial products which display some remarkably specific effects on living cells. The range of biological effects which have been shown to be susceptible to the action of the cytochalasans is already quite considerable, but perhaps the best authenticated are their interference with cytoplasmic cleavage, their inhibition of cell movement and their ability to induce cell enucleation[126]. These unusual biological properties were first encountered in 1964 when the first two cytochalasans—cytochalasin A (74) and cytochalasin B (75)—were isolated from cultures of *Helmintho-sporum dematioideum*. Their structures were determined in 1967[127, 128]. Independently, Tamm and Rothweiler[129, 130] isolated from *Phoma* species the same two compounds which they named dehydrophomin and phomin, respectively. Subsequently, further cytochalasans were isolated and identified. Cytochalasin C (76) and cytochalasin D (77) were obtained from culture filtrates of *Metarrhizium anisopliae*[131] and structures (78) and (79) were assigned[132] to cytochalasin E from *Rosellina necatrix* and cytochalasin F from *Helminthosporum dematioideum* on the basis of spectroscopic examination and chemical transformations. The Zygosporins D–G, isolated by Japanese workers from fermentations of *Zygosporium masonii*, have been shown to be variants on the cytochalasin D structure and Zygosporin A to be identical with cytochalasin C[133, 134]. The cytochalasins are related by certain common structural features and also very probably by a common biosynthetic derivation[132, 135] from L-phenylalanine, acetate and malonate units, and L-methionine. Much of this earlier work has been reviewed by Binder and

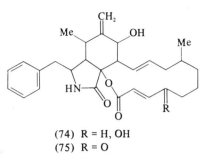

(74) R = H, OH
(75) R = O

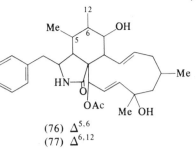

(76) $\Delta^{5,6}$
(77) $\Delta^{6,12}$

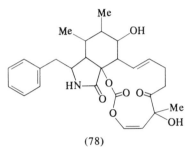

(78)

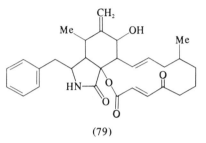

(79)

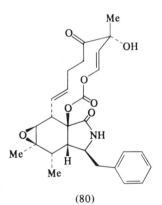

(80)

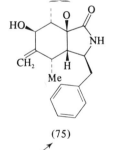

(75)

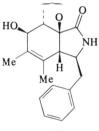

(82)

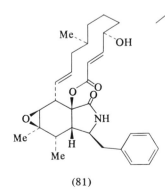

(81)

Tamm[136] and more recent investigations have added to the list of cyto-chalasans and have also led to a revision of the structures proposed earlier[132] for cytochalasins E (78) and F (79).

Buchi and his collaborators[137] have isolated a toxin from the mould *Aspergillus clavatus* grown on rice and have established its identity with cytochalasin E[132]. Examination of its spectroscopic properties and x-ray analysis of the silver ion complex showed it to possess the stereostructure

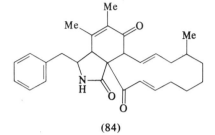

(83)   (84)

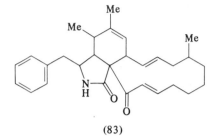

(85)   (77)

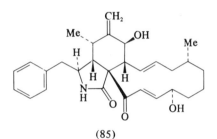

(87)

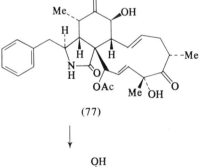

(86)

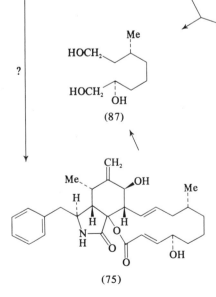

(75)

(80), in contrast to the structure (78) proposed by the I.C.I. group[132]. By analogy, the structure of cytochalasin F must also be revised to (81). The previous structural proposals for these cytochalasans were based in part on the products obtained after mild acid treatment. Thus cytochalasin F gave cytochalasin B (75) and the isomer (82) and these changes, which were previously postulated to involve a hydride shift, are unexceptional and readily accommodated by the epoxide structures.

From cultures of a *Phoma* species, three further cytochalasans with the same carbon skeleton—proxiphomin (83), protophomin (84) and deoxaphomin (85)—have been obtained recently by Binder and Tamm[138, 139]. Structures were assigned on the basis of spectroscopic data and chemical transformations. Thus deoxaphomin (85) gave the products (86) and (87) on ozonolysis and subsequent treatment with sodium borohydride, and these were identical with products obtained analogously from cytochalasin D (77) and cytochalasin B (75), respectively. Binder and Tamm speculated on the possible position of the [13]-cytochalasans such as deoxaphomin (85) as biogenetic precursors of the 24-oxa-[14]-cytochalasans such as phomin or cytochalasin B (75). However, the attempted Baeyer–Villiger oxidation of deoxaphomin to give phomin (75) was inconclusive, although a very small quantity of material with chromatographic properties identical to (75) was obtained.

Japanese workers[140] have obtained cytotoxic metabolites related to the cytochalasans but based on the aromatic amino acid L-tryptophan from *Chaetomium globosum*. Chaetoglobosin A (88) and B (89) were thus obtained and assigned the accompanying structures on the basis of spectroscopic data. Further proof of the structural relationship between the two metabolites and of the presence of an epoxide ring in (88) was obtained by transformation of chaetoglobosin A (88) to its isomer (89) by treatment with triethylamine

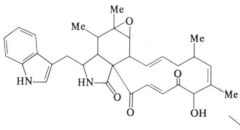

(88)

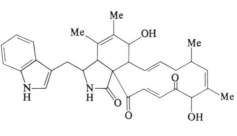

(89)

or by standing in chloroform. The change is analogous to that observed by Turner and his group[132] with cytochalasins E and F.

5.4.2 Marine natural products

The fascinating series of halogenated secondary metabolites which are currently being isolated from a range of marine organisms continue to excite the imagination of the natural products chemist. This work has focused attention particularly on the biosynthetic relationships between the various groups of organic compounds from marine organisms and has posed questions of whether fundamentally new biosynthetic pathways exist to these compounds or whether they are formed broadly speaking by variations in the routes of metabolism which are operative in terrestial organisms. The answer to these questions are awaited with interest, but meanwhile it is useful to summarise some of the more recent structural work and to note how insight into the biogenetic questions is being gained. A book[141] published recently gives a comprehensive review of much of the work in this field until early 1972.

Marine algae of the genus *Laurencia*, family Rhodomelaceae, have proved to be a very rich source of halogenated natural products. Secondary metabolites of three general types have already been isolated—sesquiterpene aromatics such as laurinterol (90)[142], spiro-fused sesquiterpenes typified by pacifenol (105)[143] and non-terpenoid C_{15} compounds of which laureatin (99)[144] is a good example. The isolation of two further types has recently been disclosed. Oppositol (91) was obtained[145] (0.1%) from *Laurencia subopposita* and the structure of this brominated sesquiterpene alcohol was deduced by x-ray analysis and by the isolation of the indan (92) after treatment with toluene-*p*-sulphonic acid. The furan derivative furocaespitane (93) was similarly isolated[146] as a minor constituent of the seaweed *Laurencia caespitosa*.

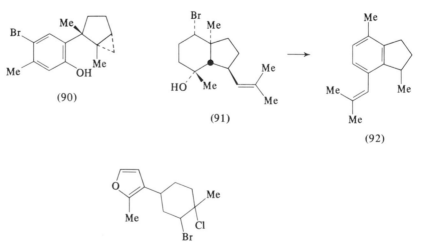

(90)   (91)   (92)

(93)

Laurefucin, a metabolite of *Laurencia nipponica*, was assigned in earlier work the structure (94) on the basis of chemical and spectroscopic evidence[147]. Reinterpretation of this evidence[148], however, led to the revised structure (95) for laurefucin, which was confirmed by x-ray analysis. Treatment of hexahydrolaurefucin with thionyl bromide gave a dibromo derivative (96), which when treated with zinc and acetic acid gave an unsaturated glycol (98)—an optical antipode of the product (97) obtained analogously from both laureatin (99) and isolaureatin (100). The same unsaturated diol was also obtained similarly from isoprelaurefucin[149] (0.001% ; 101), a component

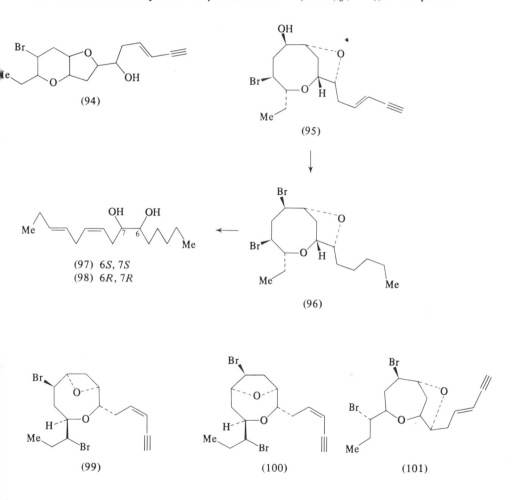

of the essential oil of *Laurencia nipponica*. The structures of both (99) and (100) have been established[150, 151] previously on the basis of their chemical transformations, the inter-relationship with laurencin (94)[152, 153] and an x-ray study of isolaureatin. Furthermore, the structural, stereochemical and biogenetic relationship between (99) and (100) was underlined by the acid catalysed rearrangement of (99) to (100).

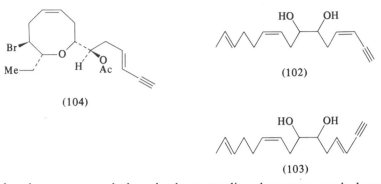

(104)

(102)

(103)

It has been presumed that the bromocyclic ether compounds laureatin (99), isolaureatin (100), laurefucin (95), laurencin (104) and isoprelaurefucin (101), isolated from the same species *Laurencia nipponica*, would be formed from a common precursor. Suggestions as to the nature of this intermediate have included hexadeca-4,7,10,13-tetraenoic acid and the laurediols (102) and (103). Reinforcing this latter proposal, Kurosawa, Fukuzawa and Irie[154] have presented evidence that the *cis*- and *trans*-laurediols (102) and (103) do indeed occur as mixtures of optical isomers in *Laurencia nipponica*.

Three further examples of the halogenated spiro-fused sesquiterpene class of natural product isolated from *Laurencia* species are johnstonol (106)[155], prepacifenol (107)[156] and caespitol (108)[157]. Like that of pacifenol (105)[143] the structures of these metabolites is readily accommodated from the biogenetic

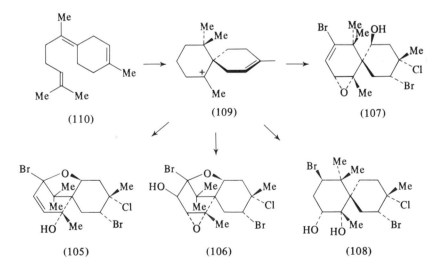

(110)

(109)

(107)

(105)

(106)

(108)

standpoint. Thus all can be formally derived from the ion (109)—a cyclisation product of γ-bisabolene (110) which has been proposed as their common precursor.

The most highly halogenated naturally occurring substance found to date is the terpene (111)[158]. Analysis of the sea hare (*Aplysia californica*) digestive

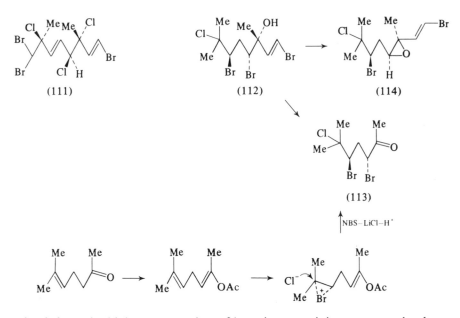

gland showed a high concentration of bromine-containing compounds whose presence has been attributed to the metabolism of halogenated products found in marine algae ingested by the sea hare. In addition to (111) the chlorobromoterpene (112) has also been characterised as a product of catabolism of the sea hare[159]. Spectral data were most easily accommodated by a structure such as (112), but the positions of the chlorine and two bromine atoms could not be assigned. Oxidation of (112) with Jones' reagent gave the ketone (113) whose structure was confirmed by synthesis. The deduction of the structure of the synthetic ketone followed on the assumption that the chloride ion underwent addition to the intermediate bromonium ion at the most highly substituted carbon atom. Comparison of the ^{1}H n.m.r. of the ketone (113) and its tribromo analogue supported this assignment. The position of the bromine at C-4 was also indicated by the formation of the epoxide (114; relative stereochemistry) with base and the loss of bromide ion. ^{1}H N.m.r. (220 MHz) analysis using lanthanide shift reagents allowed a full determination of the topology of the epoxide and hence of the original natural product[160]. The ^{1}H n.m.r. spectra of all compounds in this

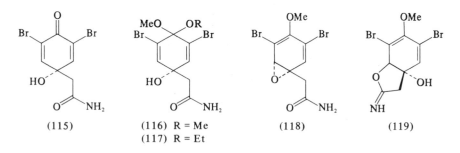

series indicated that the bulk of the halogen atoms hold the terpenoid molecules in a rigid conformation.

A cautionary note attaches finally to studies of some metabolites from sponges of genus *Verongia*. In earlier work[161] the two closely related compounds (115) and (116) were obtained from *Verongia* species and may be considered to be derived from dibromotyrosine. Faulkner and Anderson[162] have now isolated the mixed acetal (117) from an undefined species of *Verongia* and they suggested that (115), (116) and the mixed acetal might all be formed from a common precursor by the addition of solvent during the extraction process. The arene oxide (118) and the imino ether (119) were considered as possible precursors.

5.4.3   Miscellaneous new structures

As authors seek to catch the reader's eye in the rapidly proliferating scientific literature, the word novel* has become much used and abused. It can however be correctly employed to describe several of the compounds described below. Thus 1,3-dimethoxybenzene was unknown as a natural product until isolated and identified[163] as an odiferous constituent (0.4%) of the fruiting body of the fungus *Rhodophyllus icterimus*. Similarly, campholenic aldehyde (120) has been known for over 60 years as a product of photolysis of camphor but its natural occurrence along with its *trans*-epoxide (121) and the corresponding alcohol (122) and acetate (123) was reported[164] for the first time in

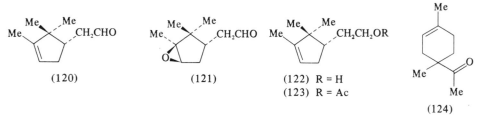

(120)   (121)   (122) R = H   (124)
                            (123) R = Ac

1972. These compounds represent the first examples in nature of the 2,2,3-trimethyl-1-ethylcyclohexane (campholenyl) skeleton. Thomas[165] has also reported the isolation of the ketone (124) from the fruit oil of *Juniperus communis*, which is a further example of a monoterpenoid of a type previously not reported in nature.

Kneifel and Bayer[166] have obtained the first naturally occurring organovanadium compound—amavadine—from fly agaric (*Amanita muscaria*). Amavadine is a pale blue compound containing vanadium (molecular weight 415) and e.s.r. and i.r. measurements indicated that the vanadium was quadrivalent in the form of a $>$V$=$O group. With acid, amavadine gave mainly L-alanine upon hydrolytic cleavage and the same amino acid, sodium pyruvate and acetaldehyde were detected after alkali treatment. Reductive

---

* *Concise Oxford Dictionary:* 'of new kind or nature, strange, hitherto unknown'.

cleavage gave αα'-iminodipropionic acid (126), but evidence of the true nature of the organic chelating agents around the vanadium atom came first from the e.s.r. detection of a nitroxyl radical on oxidation of the amavadine in alkaline media. It was finally found possible to remove the organic ligands intact by methanolysis (methanol–sulphuric acid) of amavadine, which gave dimethyl *N*-hydroxy-αα'-iminodipropionate (127). The structure (125) of the natural organovanadium compound was confirmed by a synthesis of a

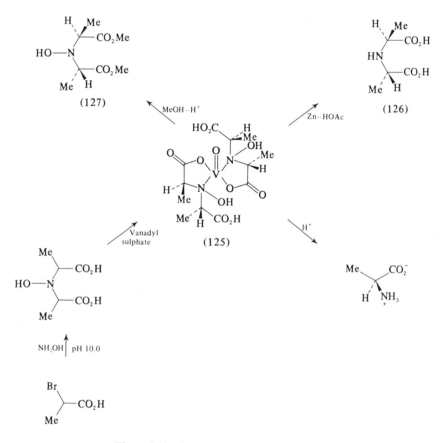

Figure 5.10 Some reactions of amavadine[166]

complex (Figure 5.10) which differed only in the chirality at the four asymmetric centres. In the natural product these four centres exist in the L-configuration.

Three new heterocyclic bases have been isolated for the first time as natural products. A new cytokinin from *Populus robusta* leaves was shown by mass spectrometry[167] and comparison with a synthetic specimen to be 6-(*o*-hydroxybenzylamino)9-β-D-ribofuranosyl purine. This represents the first isolation of a natural cytokinin with an aromatic side chain. Whilst numerous halogen containing compounds have been isolated from marine organisms, the finding of 5-chlorocytosine (128) in the DNA hydrolysate of salmon sperm represents the first case of the detection in nature of a halogenated nucleic acid constituent[168]. Its biological significance cannot yet be assessed,

(128)

but it may arise in some way simply from the high chloride concentration in the marine environment. Also isolated from the salmon sperm was 5-chloro-deoxycytidine.

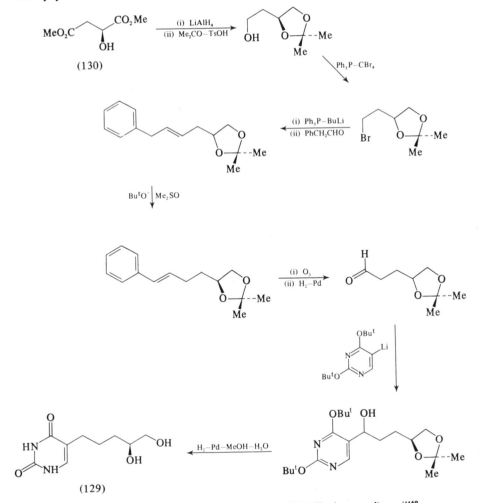

Figure 5.11   Structure and synthesis of (+)-5-(4′,5′-dihydropentyl)uracil[169]

Nakanishi and Hayashi[169] used an extremely neat synthesis (Figure 5.11) to establish the structure and absolute stereochemistry of (+)-5-(4',5'-dihydroxypentyl)uracil (129), the modified base which replaces thymine in bacteriophage SP-15 DNA of *Bacillus subtilis*. The synthesis, commencing with (S)-dimethyl malate (130), allowed the S configuration to be assigned to the chiral centre in (129). This is the sole example to date of a DNA base containing a chiral side chain.

References

1. Perkin, W. H. (1904). *J. Chem. Soc.*, 416, 654
2. Woodward, R. B. (1956). In *Perspectives in Organic Chemistry*, 125 (A. R. Todd, editor) (London and New York: Interscience)
3. Wuonola, M. A. and Woodward, R. B. (1973). *J. Amer. Chem. Soc.*, 95, 284, 5098
4. Arndt, R. R., Eggers, S. H. and Jordaan, A. (1969). *Tetrahedron*, 25, 2767
5. Robinson, R. (1974). *Chem. Brit.*, 10, 57
6. Cushley, R. J., Anderson, D. R., Lipsky, S. R., Sykes, R. J. and Wasserman, H. (1971). *J. Amer. Chem. Soc.*, 93, 6284
7. Cushley, R. J., Lipsky, S. R., Wasserman, H., Peverada, P. and Shaw, C. K. (1973). *J. Amer. Chem. Soc.*, 95, 6874
8. Fuhrer, H. (1973). *Helv. Chim. Acta*, 56, 2377
9. Milavetz, B., Kakinuma, K., Rinehart, K. L., Rolls, J. P. and Haak, W. J. (1973). *J. Amer. Chem. Soc.*, 95, 5793
10. Knoll, W. M. J., Huxtable, R. J. and Rinehart, K. L. (1973). *J. Amer. Chem. Soc.*, 95, 2703
11. Tanabe, M. and Seto, H. (1970). *J. Org. Chem.*, 35, 2087; *Biochemistry*, 9, 4851
12. Tanabe, M., Seto, H. and Johnson, L. F. (1970). *J. Amer. Chem. Soc.*, 92, 2157
13. Scott, A. I., Townsend, C. A., Okada, K., Kajiwara, M., Whitman, P. J. and Cushley, R. J. (1972). *J. Amer. Chem. Soc.*, 94, 826, 8267
14. Battersby, A. R., Gibson, K. H., McDonald, E., Mander, L. N., Moron, J. and Nixon, L. N. (1973). *J. Chem. Soc. Chem. Commun.*, 768
15. Battersby, A. R., Hunt, E. and McDonald, E. (1973). *J. Chem. Soc. Chem. Commun.*, 442
16. Baldwin, J. E., Loliger, J., Rastelter, W., Neuss, N., Huckstep, L. L. and De La Higuera, N. (1973). *J. Amer. Chem. Soc.*, 95, 3796
17. Kluender, H., Bradley, C. H., Sih, C. J., Fawcett, P. and Abraham, E. P. (1973). *J. Amer. Chem. Soc.*, 95, 6149
18. Scott, A. I., Townsend, C. A. and Cushley, R. J. (1973). *J. Amer. Chem. Soc.*, 95, 5759
19. Bray, R. C. and Shemin, D. (1963). *J. Biol. Chem.*, 238, 1501
20. Edmond, H., Crowfoot-Hodgkin, D. (1972). *J. Chem. Soc. Perkin Trans. II*, 605
21. Bonnett, R., Godfrey, J. M., Mata, V. B., Scopes, P. M. and Thomas, R. N. (1973). *J. Chem. Soc. Perkin Trans. I*, 252
22. Dalling, D. K. and Grant, D. M. (1972). *J. Amer. Chem. Soc.*, 94, 5318
23. Archer, R. A., Cooper, R. D. G., Demarco, P. V. and Johnson, L. R. F. (1970). *Chem. Commun.*, 1291
24. Morton, J. B., Long, R. C., Daniels, P. J. L., Trach, R. W. and Goldstein, J. H. (1973). *J. Amer. Chem. Soc.*, 95, 7464
25. Usui, T., Kobayashi, M., Yamuoka, N., Matsuda, K., Tuzimura, K., Sugiyama, H. and Seto, S. (1973). *Tetrahedron Lett.*, 3397
26. Pugmire, R. J., Grant, D. M., Townsend, L. B. and Robins, R. K. (1973). *J. Amer. Chem. Soc.*, 95, 2791
27. Ellis, P. D., Dunlap, P. R., Pollard, A. L., Seidman, K. and Cardini, A. D. (1973). *J. Amer. Chem. Soc.*, 95, 4398
28. Leibfritz, D. and Roberts, J. D. (1973). *J. Amer. Chem. Soc.*, 95, 4996
29. Muller, G. and Phillipsborn, W. (1973). *Helv. Chim. Acta*, 56, 2680
30. Wenkert, E., Cochran, D. W., Hagaman, E. W., Schell, F. M., Neuss, N., Katner,

A. S., Potier, P., Kan, C., Plat, M., Koch, M., Mehri, H., Poisson, J., Kunesch, N. and Rolland, Y. (1973). *J. Amer. Chem. Soc.*, **95**, 4990

31. Needham, T. E., Mataviyoff, N. A., Walker, T. E. and Hogenkamp, H. P. C. (1973). *J. Amer. Chem. Soc.*, **95**, 5019
32. Schweizer, M. P., Banta, E. B., Witowski, J. T. and Robins, R. K. (1973). *J. Amer. Chem. Soc.*, **95**, 3770
33. Lapper, R. D., Mantsch, H. H. and Smith, I. C. P. (1973). *J. Amer. Chem. Soc.*, **95**, 2878, 2880
34. Hall, L. D. and Johnson, L. F. (1969). *Chem. Commun.*, 509
35. Perlin, A. S. and Casu, B. (1969). *Tetrahedron Lett.*, 2921
36. Perlin, A. S., Casu, B. and Koch, H. J. (1970). *Can. J. Chem.*, **48**, 2596
37. Dorman, D. E. and Roberts, J. D. (1970). *J. Amer. Chem. Soc.*, **91**, 1355
38. Binkley, W. W., Horton, D., Bhacca, N. S. and Wander, J. C. (1972). *Carbohydrate Res.*, **23**, 301
39. Breitmaier, E., Voelter, W., Jung, G. and Tanzer, C. (1971). *Chem. Ber.*, **104**, 1147
40. Bock, K., Lundt, I. and Pedersen, C. (1973). *Tetrahedron Lett.*, 1037
41. Nakanishi, K., Gullo, V. P., Miura, I., Govindachari, T. R. and Viswanathan, N. (1973). *J. Amer. Chem. Soc.*, **95**, 6473
42. Mussini, P., Orsini, F., Pelizzoni, F., Buckwalter, B. L. and Wenkert, E. (1973). *Tetrahedron Lett.*, 4849
43. Ferrari, M., Pelizzoni, F. and Ferrari, G. (1972). *Phytochemistry*, **10**, 3267
44. Wenkert, E. and Buckwalter, B. L. (1972). *J. Amer. Chem. Soc.*, **94**, 4367
45. Jacques, D., Haslam, E., Bedford, G. R. and Greatbanks, D. (1973). *J. Chem. Soc. Chem. Commun.*, 518
46. Schilling, G., Weinges, K., Muller, O. and Mayer, W. (1973). *Annalen*, 1471
47. Wilson, J. M. (1973). In *Structure Determinations in Organic Chemistry* (MTP International Review of Science, Chemistry Series One), (W. D. Ollis, editor) (Butterworths: London), p. 44
48. Field, F. H. and Munson, M. S. B. (1966). *J. Amer. Chem. Soc.*, **88**, 2621
49. Field, F. H. (1968). *Accounts Chem. Res.*, **1**, 42
50. Clark-Lewis, J. W., Harwood, C. N., Lacey, M. J. and Shannon, J. S. (1973). *Aust. J. Chem.*, **26**, 1577
51. Kingston, D. G. I. and Fales, H. M. (1973). *Tetrahedron*, **29**, 4083
52. Clark-Lewis, J. W. (1968). *Aust. J. Chem.*, **21**, 2059, 3025
53. Kingston, D. G. I. (1971). *Tetrahedron*, **27**, 2691
54. Mitscher, L. A., Showalter, H. D. H. and Foltz, R. L. (1972). *J. Chem. Soc. Chem. Commun.*, 796
55. Fales, H. M., Lloyd, H. A. and Milne, G. W. A. (1970). *J. Amer. Chem. Soc.*, **92**, 1590
56. Michnowicz, J. and Munson, B. (1972). *Org. Mass. Spectrom.*, **6**, 765
57. Fales, H. M., Milne, G. W. A. and Nicholson, R. S. (1971). *Anal. Chem.*, **43**, 1785
58. Longevialle, P., Milne, G. W. A. and Fales, H. M. (1973). *J. Amer. Chem. Soc.*, **95**, 666
59. Arsenault, G. P., Althaus, J. R. and Divekar, P. V. (1969). *Chem. Commun.*, 1414
60. Sen Gupta, R., Chandran, R. R. and Divekar, P. V. (1966). *Indian J. Exp. Biol.*, **4**, 152
61. Stevens, J. D. and Fletcher, H. G. (1968). *J. Org. Chem.*, **33**, 1799
62. McMurry, P. and Abe, K. (1973). *J. Amer. Chem. Soc.*, **95**, 5824
63. Okuma, K., Addicott, F. T., Smith, O. E. and Thiessen, W. E. (1965). *Tetrahedron Lett.*, 2529
64. Cornforth, J. W., Milborrow, B. V., Ryback, G. and Wareing, F. P. (1965). *Nature*, **205**, 1269
65. Cornforth, J. W., Milborrow, B. V. and Ryback, G. (1965). *Nature*, **206**, 715
66. Cornforth, J. W., Milborrow, B. V. and Ryback, G. (1966). *Nature*, **210**, 627
67. Cornforth, J. W., Draper, W., Milborrow, B. V. and Ryback, G. (1967). *Chem. Commun.*, 114
68. Mills, J. A. (1952). *J. Chem. Soc.*, 4976
69. Burden, R. S. and Taylor, H. F. (1970). *Tetrahedron Lett.*, 4071
70. Bartlett, L., Klyne, W., Mose, W. P., Scopes, P. M., Galasko, G., Mallams, A. K., Weedon, B. C. L., Szabolos, J. and Toth, G. (1969). *J. Chem. Soc. C*, 2527
71. Isoe, S., Hyeon, S. B., Katsumura, S. and Sakau, S. (1972). *Tetrahedron Lett.*, 2517
72. Oritani, T. and Yamashita, K. (1972). *Tetrahedron Lett.*, 2521
73. Harada, N. (1973). *J. Amer. Chem. Soc.*, **95**, 240

74. Ryback, G. (1972). *J. Chem. Soc. Chem. Commun.*, 1190
75. Horn, D. H. S. and Pretorius, Y. Y. (1954). *J. Chem. Soc.*, 1460
76. Koreeda, M., Weiss, G. and Nakanishi, K. (1973). *J. Amer. Chem. Soc.*, **95**, 239
77. Cahn, R. S., Ingold, C. K. and Prelog, V. (1956). *Experientia*, **12**, 81
78. Cahn, R. S., Ingold, C. K. and Prelog, V. (1966). *Angew. Chem. Int. Ed. Engl.*, **5**, 385
79. Herz, W., Bhat, S. V. and Hall, A. L. (1970). *J. Org. Chem.*, **35**, 1100
80. Stocklin, W., Waddell, T. G. and Geismann, T. A. (1970). *Tetrahedron*, **26**, 2397
81. Sykora, V. and Romanuk, M. (1957). *Coll. Czech, Chem. Commun.*, 1909.
82. Sundararaman, P., McEwen, R. S. and Herz, W. (1973). *Tetrahedron Lett.*, 3809
83. Chickamatsu, H. and Herz, W. (1973). *J. Org. Chem.*, **38**, 585
84. McPhail, A. T., Luhan, P. A., Lee, K.-H., Furukawa, H., Meek, R., Piantadosi, C. and Shingu, T. (1973). *Tetrahedron Lett.*, 4087
85. Furukawa, H., Lee, K.-H., Meek, R., Piantadosi, C. and Shingu, T. (1973). *J. Org. Chem.*, **38**, 1722
86. Gibson, D. T., Hensley, M., Yoshioka, H. and Mabry, T. J. (1970). *Biochemistry*, **9**, 1626
87. Jerina, D. M., Daly, J. W., Jeffrey, A. M. and Gibson, D. T. (1971). *Arch. Biochem. Biophys.*, **142**, 394
88. Ziffer, H., Jerina, D. M., Gibson, D. T. and Kobal, V. M. (1973). *J. Amer. Chem. Soc.*, **95**, 4048
89. Kobal, V. M., Gibson, D. T., Davis, R. E. and Garza, A. (1973). *J. Amer. Chem. Soc.*, **95**, 4420
90. Nakanishi, K. and Harada, N. (1972). *Accounts Chem. Res.*, **5**, 257
91. Abraham, E. P. and Newton, G. G. F. (1961). *Endeavour*, **20**, 92
92. Abraham, E. P. and Newton, G. G. F. (1961). *Biochem. J.*, **79**, 377
93. Crowfoot-Hodgkin, D. and Maslen, E. N. (1961). *Biochem. J.*, **79**, 393
94. Strominger, J. L. and Tipper, D. J. (1965). *Amer. J. Med.*, **39**, 708
95. Nagarajan, R., Boeck, L. D., Gorman, M., Hammill, R. L., Higgins, C. E., Hoehn, M. M., Stark, W. M., Whitney, J. G. (1971). *J. Amer. Chem. Soc.*, **93**, 2308
96. Nagarajan, R. and Spry, D. O. (1971). *J. Amer. Chem. Soc.*, **93**, 2310
97. Cama, L. D., Leanza, W. J., Beattie, T. R. and Christensen, B. G. (1972). *J. Amer. Chem. Soc.*, **94**, 1408
98. Koppel, G. A. and Koehler, R. E. (1973). *Tetrahedron Lett.*, 1943
99. Ratcliffe, R. W. and Christensen, B. G. (1973). *Tetrahedron Lett.*, 4649, 4653
100. Cama, L. D. and Christensen, B. G. (1973). *Tetrahedron Lett.*, 3505
101. Spitzer, W. A. and Goodson, T. (1973). *Tetrahedron Lett.*, 273
102. Baldwin, J. E., Urban, F. J., Cooper, R. D. G. and Jose, F. L. (1973). *J. Amer. Chem. Soc.*, **95**, 2401
103. Koppel, G. A. and Koehler, R. E. (1973). *J. Amer. Chem. Soc.*, **95**, 2403
104. Slusarchyk, W. A., Applegate, H. E., Funke, P., Koster, W., Puar, M. S., Young, M. and Dolfini, J. E. (1973). *J. Org. Chem.*, **38**, 943
105. Shemyakin, M. M., Vinogradova, E. I., Feigina, M. Y., Aldanova, N. A., Loginova, N. F., Ryabora, I. D. and Pavlenko, I. A. (1965). *Experientia*, **21**, 548
106. Dominguez, J., Dunitz, J. D., Gerlach, H. and Prelog, V. (1962). *Helv. Chim. Acta*, **45**, 129
107. Eisenman, G. (1968). *Fed. Proc.*, **27**, 1249
108. Moore, C. and Pressman, B. (1964). *Biochem. Biophys. Res. Commun.*, **15**, 562
109. Pedersen, C. J. (1967). *J. Amer. Chem. Soc.*, **89**, 7017
110. Pedersen, C. J. and Frensdorff, H. K. (1972). *Angew. Chem. Int. Ed. Engl.*, **11**, 16
111. Prestegard, J. H. and Chan, S. I. (1970). *J. Amer. Chem. Soc.*, **92**, 4440
112. Kilbourn, B. T., Dunitz, J. D., Pioda, L. A. R. and Simon, W. (1967). *J. Mol. Biol.*, **30**, 559
113. Pinkerton, M. and Steinrauf, L. (1970). *J. Mol. Biol.*, **49**, 533
114. Lutz, W. K., Winkler, F. K. and Dunitz, J. D. (1971). *Helv. Chim. Acta*, **54**, 1103
115. Kubota, T. and Matsutani, S. (1970). *J. Chem. Soc. C*, 695
116. Bissell, E. C. and Paul, I. C. (1972). *J. Chem. Soc. Chem. Commun.*, 967
117. Alleaume, M. and Hickle, D. (1972). *J. Chem. Soc. Chem. Commun.*, 175
118. Czerwinski, E. W. and Steinrauf, L. K. (1971). *Biochem. Biophys. Res. Commun.*, **45**, 1284
119. Blount, J. F. and Westley, J. W. (1971). *Chem. Commun.*, 927

120. Kinashi, H., Otake, N., Yonehara, H., Sato, S. and Saito, Y. (1973). *Tetrahedron Lett.*, 4955
121. Jones, N. D., Chaney, M. O., Chamberlin, J. W., Hamill, R. L. and Chen, S. (1973). *J. Amer. Chem. Soc.*, **95**, 3399
122. Alpha, S. R. and Brady, A. H. (1973). *J. Amer. Chem. Soc.*, **95**, 7043
123. Ivanov, V. T., Miroshnikov, A. I., Abdullaev, N. D., Senyavina, L. B., Arkhipova, S. F., Uvarova, N. N., Khalilulina, K. Kh., Bystrov, V. F. and Orchimnikov, Y. A. (1971). *Biochem. Biophys. Res. Commun.*, **42**, 654
124. Urry, D. W., Glickson, J. D., Mayers, D. F. and Haider, J. (1972). *Biochemistry*, **11**, 487
125. Sarges, R. and Witkop, B. (1965). *J. Amer. Chem. Soc.*, **87**, 2011
126. Carter, S. B. (1972). *Endeavour*, **31**, 77
127. Aldridge, D. C., Armstrong, J. J., Speake, R. N. and Turner, W. B. (1967). *Chem. Commun.*, 26
128. Aldridge, D. C., Armstrong, J. J., Speake, R. N. and Turner, W. B. (1967). *J. Chem. Soc. C*, 1667
129. Rothweiler, W. and Tamm, Ch. (1966). *Experientia*, **22**, 750
130. Rothweiler, W. and Tamm, Ch. (1970). *Helv. Chim. Acta*, **53**, 696
131. Aldridge, D. C. and Turner, W. B. (1969). *J. Chem. Soc. C*, 923
132. Aldridge, D. C., Burrows, B. F. and Turner, W. B. (1972). *Chem. Commun.*, 148
133. Minato, H. and Matsumoto, M. (1970). *J. Chem. Soc. C*, 38
134. Minato, H. and Katayama, T. (1970). *J. Chem. Soc. C*, 45
135. Binder, M., Kiechel, J. R. and Tamm, Ch. (1970). *Helv. Chim. Acta*, 53, 1797
136. Binder, M. and Tamm, Ch. (1973). *Angew. Chem. Int. Ed. Engl.*, **12**, 370
137. Büchi, G., Kitaura, Y., Yuan, S.-S., Wright, H. E., Clardy, J., Demain, A. L., Glinsukan, T., Hunt, N. and Wogan, G. N. (1973). *J. Amer. Chem. Soc.*, **95**, 5423
138. Binder, M. and Tamm, Ch. (1973). *Helv. Chim. Acta*, **56**, 966
139. Binder, M. and Tamm, Ch. (1973). *Helv. Chim. Acta*, **56**, 2387
140. Sekita, S., Yoshihira, K., Natori, S. and Kuwano, H. (1973). *Tetrahedron Lett.*, 2109
141. Scheuer, P. J. (1973). *Chemistry of Marine Natural Products*, (London and New York: Academic Press)
142. Irie, T., Suzuki, M., Kurosawa, E. and Masamune, T. (1970). *Tetrahedron*, **26**, 3271
143. Sims, J. J., Fenical, W., Wing, R. M. and Radlick, P. (1971). *J. Amer. Chem. Soc.*, **93**, 3774
144. Irie, T., Izawa, M. and Kurosawa, E. (1970). *Tetrahedron*, **26**, 851
145. Hall, S. S., Faulkner, D. J., Fayos, J., Clardy, J. (1973). *J. Amer. Chem. Soc.*, **95**, 7187
146. Gonzalez, A. G., Darias, J. and Martin, J. D. (1973). *Tetrahedron Lett.*, 3625
147. Fukuzawa, A., Kurosawa, E. and Irie, T. (1972). *Tetrahedron Lett.*, 3
148. Furusaki, A., Kurosawa, E., Fukuzawa, A. and Irie, T. (1973). *Tetrahedron Lett.*, 4579
149. Kurosawa, E., Fukuzawa, A. and Irie, T. (1973). *Tetrahedron Lett.*, 4135
150. Fukuzawa, A., Kurosawa, E. and Irie, T. (1972). *J. Org. Chem.*, **37**, 680
151. Kurosawa, E., Furusaki, A., Izawa, M., Fukuzawi, A. and Irie, T. (1973). *Tetrahedron Lett.*, 3857
152. Irie, T., Suzuki, M. and Masamune, T. (1968). *Tetrahedron*, **24**, 4193
153. Cameron, A. F., Cheung, K. K., Ferguson, G. and Robertson, J. M. (1969). *J. Chem. Soc. B*, 559
154. Kurosawa, E., Fukuzawa, A. and Irie, T. (1972). *Tetrahedron Lett.*, 2121
155. Sims, J. J., Fenical, W., Wing, R. M. and Radlick, P. (1972). *Tetrahedron Lett.*, 195
156. Sims, J. J., Fenical, W., Wing, R. M. and Radlick, P. (1973). *J. Amer. Chem. Soc.*, **95**, 972
157. Gonzalez, A. G., Darias, J., Martin, J. D. (1973). *Tetrahedron Lett.*, 2381
158. Faulkner, J. D., Stallard, M. O., Fayos, J. and Clardy, J. (1973). *J. Amer. Chem. Soc.*, **95**, 3413
159. Faulkner, J. D. and Stallard, M. O. (1973). *Tetrahedron Lett.*, 1171
160. Willcott, M. R., Davis, R. E., Faulkner, D. J. and Stallard, M. O. (1973). *Tetrahedron Lett.*, 3967
161. Sharma, G. M. and Burkholder, D. R. (1967). *Tetrahedron Lett.*, 4147
162. Anderson, R. J. and Faulkner, D. J. (1973). *Tetrahedron Lett.*, 1175
163. Schmitt, J. A. and Klose, W. (1973). *Annalen*, 544
164. Thomas, A. F. (1973). *Helv. Chim. Acta*, **56**, 1800

165. Thomas, A. F. (1972). *Helv. Chim. Acta*, **55**, 815
166. Kneifel, H. and Bayer, E. (1973). *Angew. Chem. Int. Ed. Engl.*, **12**, 508
167. Horgan, R., Hewett, E. W., Purse, J. G. and Wareing, P. F. (1973). *Tetrahedron Lett.*, 2827
168. Lis, A. W., McLaughlin, R. K., McLaughlin, D. I., Davies, G. D. and Anderson, W. R. (1973). *J. Amer. Chem. Soc.*, **95**, 5789
169. Hayashi, H. and Nakanishi, K. (1973). *J. Amer. Chem. Soc.*, **95**, 4081

Index

A-204A
  structure determination, 146
Abscisic acid
  configuration and conformation, 137–140
Acenaphthene, nitro-
  mass spectrometry, 8
Acetanilide
  mass spectrometry, 14
Acetophenone
  substituted, mass spectrometry, 5
Acetylation
  aromatic compounds, gas phase, 22
Acetylene, di-iodo-
  x-ray crystallography, 107
Acids (see also Abscisic; Ascorbic; But-3-ynoic; Carboxylic; Cholic; Cyclopentanecarboxylic; Tartaric; and Terephthalic acids)
Acridine
  spectroscopy, 47
——, amino-
  spectroscopy, 48
Adrenaline
  x-ray crystallography, 119
Alcohols (see also Cyclohexanol; Phenols; Phloroglucinol; Presqualene alcohol; Promedol alcohol)
  absolute configuration determination by n.m.r., 77
  acid-catalysed dehydration, 21
Aldehydes (see also Benzaldehyde; Campholenic aldehyde; Hexanal)
Aldrin
  mass spectrometry, 29
  x-ray crystallography, 121
Alkaloids
  ^{13}C n.m.r., 129
Aliphatic compounds
  molecular ions, mass spectrometry, 15, 16
Alupent
  x-ray crystallography, 119
Amavadine
  as natural product, 156

Amines
  absolute configuration determination by n.m.r., 76
  basicities, 19
  optical purity determination by n.m.r., 76
Amino acids
  structure, mass spectrometry and, 22–25
Ammonia
  in chemical ionisation mass spectrometry, 4
Amphetamine
  x-ray crystallography, 119
Androstene
  5,6-unsaturated, x-ray crystallography, 117
Anions
  ^{13}C n.m.r., 61, 62
  production in mass spectrometry, 7, 8
Annonalide
  ^{13}C n.m.r., 131
Antamamide
  structure determination, 147
Anthracene
  complex with 1,2,4,5-tetracyanobenzene, x-ray crystallography, 106
——, perhydro-
  conformation, n.m.r., 89
Antibiotics
  macrolide, chemical ionisation mass spectrometry, 134
  mass spectrometry, 27
  structure determination, 142–152
Antibiotic K 16
  u.v. spectroscopy, 48
Apachin
  stereochemistry, 140
Aromatic compounds (see also Anthracene; Azanaphthalene; Benzene; Biphenyl; Biphenylene; Naphthalene; Phenalene; Phenanthrene; Toluene)
  molecular ions, structures, mass spectrometry, 12–15
  spectroscopy, 45–47
Aromaticity
  x-ray crystallography and, 105

L-Ascorbic acid, dehydro-
dimer, x-ray crystallography, 117
Avenaciolide
biosynthesis, ^{13}C n.m.r. and, 65
Axerophtene
spectroscopy, 44
Azanaphthalene, amino-
spectroscopy, 48
Azines
spectroscopy, 41, 42
Azo compounds
spectroscopy, 39

Barbiturates
mass spectrometry, 3
Beckmann rearrangement
solid state, x-ray crystallography and, 109
Benzaldehyde
oxime methyl ethers, mass spectrometry, 10
2,3-Benzcarbazoloquinones
spectroscopy, 48
Benzene
derivatives, spectroscopy, 45, 46
molecular ion, structure, mass spectrometry, 12
——, chloro-
ion kinetic energy spectroscopy, 14
——, p-diethynyl-
x-ray crystallography, 106
——, 1,3-dimethoxy-
as natural product, 156
——, 1,2,4,5-tetrabutyl-
conformation, x-ray crystallography, 111
——, 1,2,4,5-tetracyano-
complex with anthracene, x-ray crystallography, 106
——, 2,4,6-trinitro-
complex with 3-formylbenzothiophene, x-ray crystallography, 106
Benzenesulphon-p-aniside, p-methoxy-
x-ray crystallography, 120
Benzimidazoles
spectroscopy, 49
11 H-Benzo[a]carazoles
spectroscopy, 49
Benzocyclo-octatetraene
conformation, n.m.r., 91
Benzonitrile, 2,4,6-tribromo-
x-ray crystallography, 107
——, 2,4,6-trichloro-
x-ray crystallography, 107
Benzo[g]quinoline, amino-
spectroscopy, 48
Benzothiophene, 3-formyl-
complex with 2,4,6-trinitrobenzene, x-ray crystallography, 106
Benzyl chloride
molecular ions, structure, mass spectrometry, 13

Bicyclo[2.2.2]octane
derivatives, conformation, x-ray crystallography, 115
Biosynthesis
^{13}C n.m.r. and, 62–66
Biphenyl
derivatives, conformation, x-ray crystallography, 112
nuclear Overhauser effect in n.m.r., 84
——, dichloro-
ion kinetic energy spectroscopy, 14
——, tetrachloro-
ion kinetic energy spectroscopy, 15
Biphenylene
spectroscopy, 46
Bonding
intramolecular, x-ray crystallography and, 100–106
Borates, tetraphenyl-
shift reagent for sulphonium compounds, 75
Borazine, hexachloro-
conformation, x-ray crystallography, 114
Borneol
molecular geometry, 70
Botryodiplodin
chemical ionisation mass spectrometry, 136
Brucine
^{13}C n.m.r., 83
But-2-ene
reaction with ozone, mass spectrometry, 6
But-3-ynoic acid
x-ray crystallography, 108

Caespitol
structure determination, 154
Campholenic aldehyde
as natural product, 156
Carbene, phenyl-
^{13}C n.m.r. and, 66
Carbohydrates
^{13}C n.m.r., 129
spin–lattice relaxation time, 82
Carbonyl compounds (see also Aldehydes; Carboxylic acids; Ketones)
spectroscopy, 39
Carboxylic acids (see also Dicarboxylic acids) acidities, 20
Carolenalin
stereochemistry, 140
Carolenin
stereochemistry, 140
Cations
^{13}C n.m.r., 61, 62
Cephalosporins
biosynthesis, ^{13}C n.m.r. and, 63
stereochemistry, ^{13}C n.m.r., 128

Cephamycins
  structure determination, 142–145
Chaetoglobosin A
  structure determination, 151
Chaetoglobosin B
  structure determination, 151
Chemical ionisation mass spectrometry, 2–4
  natural product structure determination
    by, 132–136
Chloro compounds
  mass spectrometry, 2
Chlorophylls
  spectroscopy, 49
Chloroquine
  lanthanide shift reagents in n.m.r., 72
Chloesteryl chloride
  spin–lattice relaxation time, 83
Cholic acid
  hydrocarbon complexes, ^{13}C n.m.r., 59
——, deoxy-
  hydrocarbon complexes, ^{13}C n.m.r., 59
Chromatography (see also Gas chromato-
    graphy; Liquid chromatography)
Chromium, cycloheptatrienetricarbonyl-
  mass spectrometry, 13
Cobalomin, cyano-
  biosynthesis, ^{13}C n.m.r. and, 129
Cobalt tetraphenylporphyrin
  shift reagent in n.m.r., 75
Codeine
  ^{13}C n.m.r., 83
Configuration
  determination, coupling constants and,
    78
  natural products, determination, 137–142
  ^{13}C n.m.r. and, 60
Conformation
  determination, coupling constants and,
    78
  lanthanide shift reagents in n.m.r. and,
    73
  molecular, x-ray crystallography and,
    110–115
  natural products, determination, 137–142
  n.m.r. and, 86–93
Corrins
  ^{13}C n.m.r., 129
  spectroscopy, 49
Coupling constants, 78–81
Cyclobutane
  conformation, x-ray crystallography, 113
Cyclodecane
  conformation, n.m.r., 91
Cyclodecanone
  conformation, n.m.r., 91
Cycloheptatriene
  molecular ions, structure, mass spectro-
    metry, 13
Cyclohexene
  mass spectrometry, 6

Cyclohexa-1,3-diene
  spectroscopy, 40
Cyclohexane, t-butyl-
  derivatives, conformation, x-ray crystallo-
    graphy, 114
Cyclohexanol, t-butyl-
  cis- and trans-, mass spectrometry and, 28
Cyclohexanone
  ring inversion, n.m.r., 88
Cyclohexasilane, dodecamethyl-
  conformation, x-ray crystallography, 114
Cyclo-octane
  derivatives, conformation, n.m.r., 90
Cyclopentanecarboxylic acid, 2-diethyl-
    aminoethyl-l-phenyl-
  x-ray crystallography, 119
Cyclophanes
  conformation, n.m.r., 92
  x-ray crystallography, 113
Cyclopropane
  derivatives, structure, x-ray crystallo-
    graphy, 104
  molecular ion, mass spectrometry, 15
  visible and u.v. spectroscopy, 50
Cyclotetracosane
  conformation, n.m.r., 91
Cyclotetradecane
  conformation, x-ray crystallography, 115
Cyclotrimethylenetrinitramine
  conformation, x-ray crystallography, 114
Cytidine, 5-chlorodeoxy-
  from salmon sperm, 158
Cyctochalasans
  structure determination, 148–152
Cytosine, 5-chloro-
  in DNA hydrolysate of salmon sperm, 158

Data processing
  in mass spectrometry, 8, 9
o,p'-DDT
  x-ray crystallography, 121
p,p'-DDT
  x-ray crystallography, 121
DENDRAL system, 9
Deoxygenation
  spin–lattice relaxation times and, 84
Deuterium isotope effects
  in ^{13}C n.m.r., 58, 59
Dextrans
  ^{13}C n.m.r., 129
Dianemycin
  structure determination, 146
Diazepam
  x-ray crystallography, 118
4H-1,2-Diazepine, 3,5,7-triphenyl-
  picrate, conformation, x-ray crystallo-
    graphy, 114
Diazetidinone
  conformation, n.m.r., 87

Dibenz[b,f]azepine
    spectroscopy, 49
Dibenzocyclo-octa-1,5-diene
    conformation, n.m.r., 90
Dibenzofuran
    spectroscopy, 49
Dicarboxylic acids
    mass spectrometry, 8
Dienes
    spectroscopy, 41
Dihydrodiols
    stereochemistry, 141
Dimethyl ether
    mass spectrometry, 11
3,5-Dioxabicyclo[5.1.0]octane,   syn-8,8-di-
    chloro-4-phenyl-
    conformation, x-ray crystallography, 115
Disaccharides
    fructose-containing, mass spectrometry in
        determination of, 26
Drugs
    u.v. spectroscopy, 51
Dyes
    triphenylmethane, spectroscopy, 47
Dysprosium
    shift reagent in n.m.r., 69

Endrin
    x-ray crystallography, 121
Ephedrine
    (−)-, x-ray crystallography, 119
Erythromycin B
    chemical ionisation mass spectrometry,
        134
Esterification
    acid-catalysed gas phase, 22
Ethane
    molecular ion, mass spectrometry, 15
Ethers (see also Dimethyl ether)
Ethylene
    molecular ion, mass spectrometry, 16
——, nitro-
    spectroscopy, 41
Europium
    shift reagent in n.m.r., 69

Fenchone
    homo-enolisation, ^{13}C n.m.r., 58
Field desorption
    in mass spectrometry, 4–6
Field ionisation mass spectrometry, 4–6
Films
    stretched, polarisation measurements in,
        38
Flavonoids
    chemical ionisation mass spectroscopy,
        132
Flavothebaone
    spectroscopy, 50

Fluorescein
    spectroscopy, 49
Flourine
    proton coupling, 79
Fourier transform nuclear magnetic reson-
        ance
    ^{13}C, 57
Furazan
    spectroscopy, 49
Furocaespitane
    structure determination, 152

Gas chromatography
    mass spectrometry, 2
Geometry
    molecular, lanthanide shift reagents in
        n.m.r. and, 70
Germanium tetraphenylporphyrin
    shift reagent for n.m.r., 74
Glucose, deoxyfluoro-
    6-phosphate, mass spectrometry, 26
Glucose 6-phosphate
    mass spectrometry, 26
Glycosides, amino-
    ^{13}C n.m.r., 129
Gramicidin A
    structure determination, 147
Grisorixin
    structure determination, 146
Guanosine monophosphate
    nuclear Overhauser effect in n.m.r., 84

Halides
    ions, solvation by hydrogen bond donors,
        21
Helicenes
    nuclear Overhauser effect, 85
Helicobasidin
    biosynthesis, ^{13}C n.m.r. and, 66
Heptafulvalene
    x-ray crystallography, 105
Hesperitin
    chemical ionisation mass spectrometry,
        132
Heterocyclic compounds (see also specific
        compounds)
    spectroscopy, 47–50
Hexahelicene
    conformation, x-ray crystallography, 111
Hexanal
    mass spectrometry, 6
Hexa-1,3,5-triene
    spectroscopy, 40
——, trans-1,6-diphenyl-
    polarisation measurements, 38
Hydantoin, diphenyl-
    x-ray crystallography, 118
——, thio-
    mass spectrometry for identification of, 22

Hydrazones
  spectroscopy, 43, 44
Hydrocarbons (*see also* specific compounds)
  acidities, 20
  doubly charged ions, mass spectrometry,
    16, 17
Hydrogen bonding
  x-ray crystallography and, 110

Imidazo[1,2-*b*]pyrazole
  spectroscopy, 48
Indole, vinyl-
  u.v. spectra, 48
Inert gases
  in chemical ionisation mass spectro-
    metry, 4
Inosine, 2′,3′-isopropylidene-
  nuclear Overhauser effect in n.m.r., 84
Insecticides (*see also* Aldrin; DDT)
  x-ray crystallography, 121
Ion kinetic energy spectroscopy, 14
Ion–molecule reactions
  mass spectrometry and, 18–22
Ions (*see also* Anions; Cations)
  doubly charged, mass spectrometry, 16,
    17
  metastable, structure, mass spectrometry,
    10
  structure, mass spectrometry and, 10–12
Ion transport
  antibiotics in, 145–148
Iron phthalocyanine
  complexes, shift reagent in n.m.r., 75
Isoborneol
  molecular geometry, 70
Isodrin
  mass spectrometry, 29
Isoflavone, allyl-
  spectroscopy, 49
Isolongistrobene
  structure, 126
Isoproterenol sulphate
  x-ray crystallography, 119
Isoxazolin-5-one
  spectroscopy, 49
2-Isoxazolinium
  spectroscopy, 49
Ivambrin
  stereochemistry, 140

Johnstonol
  structure determination, 154

Ketones (*see also* Cyclodecanone; Cyclo-
    hexanone; Pyridone; 2-Quinolone;
    Tropolone; Tropone)
  cyclopropyl, spectroscopy, 39
  β-silyl αβ-unsaturated, spectroscopy, 40

Lanthanide shift reagents
  chiral, 75–78
  n.m.r., 67–74
Laser ionisation
  in mass spectrometry, 7
Laureatin
  structure determination, 152
Laurefucin
  structure determination, 153
Laurinterol
  structure determination, 152
Light scattering
  u.v. and visible spectroscopy and, 37
Liquid chromatography
  mass spectrometry and, 2
Longistrobene, dehydro-
  structure, 126

Manxine
  conformation, n.m.r., 89
Mass spectrometry, 1–33
McLafferty rearrangement
  mass spectrometry and, 14
Melamine, hexamethyl-
  conformation, x-ray crystallography, 111
Mescaline
  x-ray crystallography, 119
[7]Metacyclophane
  conformation, n.m.r., 92
Metaparacyclophanes
  conformation, n.m.r., 92, 93
Methane
  in chemical ionisation mass spectrometry,
    2
1,6-Methano[10]annulene
  substituted, structures, x-ray crystallo-
    graphy, 101
Menthone
  mass spectrometry, 5
Methylation
  aromatic compounds gas phase, 22
Molar absorption coefficient
  units, 38
Molecular complexes
  intermolecular bonding interactions and,
    106–110
Monensin
  structure determination, 146

Naphthalene
  spectroscopy, 46
——, chloro-
  ion kinetic energy spectroscopy, 14
——, 1,8-di-t-butyl-
  conformation, n.m.r., 93
——, 2-fluoro-
  carbon–fluorine coupling in, 80

Naringenin
  chemical ionisation mass spectrometry, 132
Natural products
  marine, structure determination, 152–156
  mass spectrometry in structure determination of, 26, 27
  structure determination, 125–163
Neuridine, 2′,3′-isopropylidene-
  nuclear Overhauser effect in n.m.r., 84
Nigericin
  structure determination, 146
Nitrates
  esters, spectroscopy, 39
Nitration
  aromatic compounds gas phase, 22
Nitric oxide
  in chemical ionisation mass spectrometry, 4
Nitro compounds
  aromatic, spectroscopy, 46
  mass spectrometry, 8
Nonactin
  potassium transport and, 146
Norbornene
  derivatives, long range proton–proton coupling in, 79
Nuclear magnetic resonance (see also Fourier transform nuclear magnetic resonance), 55–98
  ¹³C, 57–66
    natural products, 128–131
Nuclear Overhauser effect, 84–86
Nucleosides
  mass spectrometry, 26
  sugar ring in, conformational analysis, 78
Nucleotides
  ¹³C n.m.r., 129
  sugar ring in, conformational analysis, 78

Oligosaccharides
  fructose-containing, mass spectrometry in determination of, 26
Oospora virescens
  terpene biosynthesis in, ¹³C n.m.r. and, 62
Oppositol
  structure determination, 152
Optical purity
  determination by n.m.r., 76
Orange II
  in stretched poly(vinyl alcohol), polarisation measurements, 38
Oxamides
  spectroscopy, 41
4H-1,4-Oxazine, N-aryl-2,3-dihydro-3-oxo-
  spectroscopy, 49
Oxaziridines
  coupling constants, stereochemistry and, 80

Oxygen compounds
  proton affinities, 20
Ozone
  reaction with but-2-ene, mass spectrometry, 6

Pacifenol
  structure determination, 152
Parthemollin
  stereochemistry, 140
Penicillins
  biosynthesis, ¹³C n.m.r. and, 63
  ¹³C n.m.r., 129
  mode of action, 143
  stereochemistry, ¹³C n.m.r., 128
Pentaerythritol
  derivatives, mass spectrometry, 5
Peptides (see also Tripeptides)
  conformational analysis, 79
  structure, mass spectrometry and, 22–25
  x-ray crystallography, 116
Phenalene
  spectroscopy, 46
Phenanthrene, fluoro-
  carbon–fluorine coupling in, 80
1,10-Phenanthroline
  complex with tetracyanoquinodimethane, x-ray crystallography, 106
Phenols
  acidities, 20
Phloroglucinol, trimethyl-
  dimeric oxidation product, bond lengths in, 100
Phosphoranes
  mass spectrometry, 8
Phosphorus compounds
  carbon–phosphorus coupling constants, 80, 81
  coupling constants, 80
  structure, x-ray crystallography, 104
Photoionisation
  in mass spectrometry, 6,7
Pigments
  macrocyclic, spectroscopy, 49, 50
  spectroscopy, 44
d-Pilocarpine
  spectroscopy, 48
Polycyclic compounds (see also Aromatic compounds and specific compounds)
  spectroscopy, 46, 47
Porphyrin
  biosynthesis, ¹³C n.m.r. and, 63
  spectroscopy, 49
  type III, biosynthesis, ¹³C n.m.r., 128
Prepacifenol
  structure determination, 154
Presqualene alcohol
  stereochemistry, lanthanide shift reagents in n.m.r. and, 73

Pristimerin
^{13}C n.m.r., 83
Promedol alcohol
x-ray crystallography, 120
Propylene
molecular ion, mass spectrometry, 15
Proton affinities, 18
Pteridines
^{13}C n.m.r., 129
Purines
^{13}C n.m.r., 129
mass spectrometry, 4
——, 6-(o-hydroxybenzoamino)-9-β-D-ribo-
furanosyl-
as natural product, 157
Pyrazole, 4-nitro-
spectroscopy, 48
——, 4-nitroso-
spectroscopy, 48
Pyrazolo[5,1-b]quinazolone
spectroscopy, 48
Pyridazines
spectroscopy, 47
Pyridines
spectroscopy, 47
——, amino-
spectroscopy, 48
Pyridone
spectroscopy, 48
Pyrimidines
^{13}C n.m.r., 129
spectroscopy, 47
——, amino-
spectroscopy, 48

Quinaldine, 4-amino-
spectroscopy, 48
Quinodimethane, tetracyano-
complexes, x-ray crystallography, 106
Quinoline, amino-
spectroscopy, 48
——, 2-amino-
spectroscopy, 48
——, cis-decahydro-
conformation, n.m.r., 89
2-Quinolone, 5,6,7,8-tetrahydro-
photodimer, bond lengths in, 100
Quinones
spectroscopy, 46
Quinonoids
bridgehead methyl, conformation, n.m.r.
93
Quinoxaline, amino-
spectroscopy, 48

Reaction mechanisms
^{13}C n.m.r. and, 62–66

Rearrangements (see also Beckmann rearrangement; McLafferty rearrangement
molecular, x-ray crystallography and, 108
Retinal
all-trans-, conformation, x-ray crystallography, 118
11-cis-, conformation, x-ray crystallography, 118
spectroscopy, 44
Rhodopsin
spectroscopy, 45
Rifamycins
^{13}C n.m.r., 129
Roxburghine B
configuration determination by coupling
constants, 79

Salbutamol
x-ray crystallography, 119
Salinomycin
structure determination, 146
Self-Training Interpretive and Retrieval
System, 9
Sesquiterpenes
nuclear Overhauser effects, 85
Shift reagents (see also Lanthanide shift
reagents)
n.m.r., 67–78
Sodium alkylsulphonates
mass spectrometry, 7
Solid state reactions
x-ray crystallography and, 108
Solvation
ions, gas phase, 21
Solvent effects
heterocyclic compounds, spectroscopy
and, 49
Solvents
chiral, n.m.r. and, 75
Spectroscopy (see also Chemical ionisation
mass spectrometry; Field ionisation
mass spectrometry; Ion kinetic energy
spectroscopy; Mass spectrometry; Nuclear magnetic resonance; Ultraviolet
spectroscopy; Visible spectroscopy;
X-ray crystallography)
difference, 37
dual wavelength, 37
Spin–lattice relaxation time, 81–84
Spiro(indene-1,7'-norcaradiene)
structure, x-ray crystallography, 101
Staphylomycin S
mass spectrometry in structure determination of, 25
Stereochemistry
coupling constants and, 80
mass spectrometry and, 27–29

Stereochemistry *continued*
  visible and u.v. spectroscopy and, 50, 51
Steroids (*see also* Androstene; Cholesteryl
    chloride; Cholic acid)
  benzoic esters, polyethylene films, polar-
    isation measurements, 38
  ^{13}C n.m.r., 129
Streptovarian
  biosynthesis, ^{13}C n.m.r. and, 63
Streptovaricins
  ^{13}C n.m.r., 129
Styrenes
  derivatives, conformation, n.m.r., 88
  spectroscopy, 45
Substitution reactions
  aromatic, gas phase, 22
Sulphonium compounds
  shift reagent for, 75
Sulphur compounds
  mass spectrometry, 8

Tartaric acid, dibenzoyl-
  conformation, n.m.r., 87
Terephthalic acid, 3,6-dichloro-2,5-dihy-
    droxy-
  phase changes, x-ray crystallography, 108
Tetrabenzopentafulvalene, *trans*-1,1'-diflu-
    oro-
  fluorine–proton coupling in, 79
sym-Tetrazine, 3,6-diphenyl-
  x-ray crystallography, 106
Th 1165*a*
  x-ray crystallography, 119
Th 1179
  x-ray crystallography, 119
6a-Thiathiophthen
  structure, x-ray crystallography, 102
Thieno[3,4-*c*]thiophene, tetraphenyl-
  x-ray crystallography, 105
Thiophene
  spectroscopy, 49
Thyrotropin releasing hormone
  mass spectrometry in structure deter-
    mination of, 25
Tingenin A
  ^{13}C n.m.r., 130
Tingenin B
  ^{13}C n.m.r., 130
Toluene
  molecular ion, doubly charged, mass
    spectrometry, 17
  structure, mass spectrometry, 12, 13

Triazoles
  derivatives, structures, x-ray crystallo-
    graphy, 105
Tricyclo[3.2.1.0^{1,5}]octane, 8,8-dichloro-
  structure, x-ray crystallography, 101
Tripeptides
  mass spectrometry, 2
1,3,5-Triselenane, 2,4,6-trimethyl-
  conformation, x-ray crystallography. 114
Tropolone
  spectroscopy, 46
  x-ray crystallography, 106
——, 4-isopropyl-
  x-ray crystallography, 106
Tropone
  x-ray crystallography, 106
——, 3-azido-
  x-ray crystallography, 106
Tryptycenes
  spectroscopy, 50
——, methyl-
  conformation, n.m.r., 93

Ultraviolet spectroscopy, 35–54
Undecapentaene, 2,10-dimethyl-
  spectroscopy, 44
Uracil, (+)-5-(4',5'-dihydroxypentyl)-
  structure and synthesis, 158, 159
Urethanes
  conformation by, n.m.r., 87

Valinomycin
  potassium transport and, 146
Vanadium
  natural products containing, 156
Visible spectroscopy, 35–54
Vitamin A, anhydro-
  spectroscopy, 44
Vitamin B$_{12}$
  biosynthesis, ^{13}C n.m.r. and, 64, 128
Vitamin C
  x-ray crystallography, 116

X-206
  structure determination, 146
X-537A
  structure determination, 146
X-ray crystallography, 99–124

Ytterbium
  shift reagent in n.m.r., 69